International Trade in Water Rights

International Trade in Water Rights
The Next Step

Aline Baillat

Publishing
London • New York

Published by **IWA Publishing**
Alliance House
12 Caxton Street
London SW1H 0QS, UK
Telephone: +44 (0)20 7654 5500
Fax: +44 (0)20 654 5555
Email: publications@iwap.co.uk
Web: www.iwapublishing.com

First published 2010

Originated by The Manila Typesetting Company
Cover by designforpublishing.co.uk

British Library Cataloguing in Publication Data
A CIP catalogue record for this book is available from the British Library

Library of Congress Cataloging- in-Publication Data
A catalog record for this book is available from the Library of Congress

ISBN: 9781843393610
ISBN 10: 1843393611

To Rafael, Gabriel and Paul

Table of Contents

ISBN: 9781843393610. Published by IWA Publishing, London, UK.

Preface

I wanted to write this book for two main reasons. First I wanted to bring new ideas to the debate over international water transfers. Many things have been said and written on the 'commodification' of water resources and the application of international trade law to water resources. For me, these discussions forget to address key issues, such as the domestic and international property regimes of the water resources at stake. The book focuses more specifically on international water resources, i.e. water in international watercourses, lakes and aquifers. The fundamental question raised by international water transfers from an international river, lake or aquifer (such as the transfer from the Rhône River to Barcelona) is: what are the rights of non-riparian states to these types of water? What body of rules should be applied?

The second motivation that pushed me to write this book was my reaction to recent studies that rely upon a misleading application of the Coase Theorem to water resources and propose to use market-based instruments to help resolve disputes over international watercourses. I present an alternative understanding of these issues and develop the key challenges of international trade in water rights.

To address these two topics, I organized my reasoning around the following research question: what are the consequences of the recognition of water as an economic good in domestic water policies on international watercourses management? My academic background is in political science and international relations. When I first started to study water resources, it was as a source of conflict or cooperation among riparian states. But I soon realized that to understand water management in international river basins, it was necessary to go within the states and study domestic water policy. One of the strengths of this book is precisely

ISBN: 9781843393610. Published by IWA Publishing, London, UK.

the way comparative water policy and international relations have been brought together.

In the first part of the book, I argue that the results of domestic policies recognizing water as an economic good depend on the way water resources' property regime has been defined and implemented. This part may be particularly relevant for all those interested in the major water markets systems of the world.

The second part of the book demonstrates that, at the international level, international watercourses are usually not governed by a clear property regime. This is shown by the increasing interference of international trade law with international water law.

Finally, the third part illustrates that without a strong institutional framework defining the property regime of the shared watercourse, market-based instruments will not ease the resolution of disputes among riparian states. Policy recommendations are drawn from a comparison of two up-to-date interstate cases: the Colorado River Basin and the Murray-Darling Basin. As the experiences studied in these two cases are very recent, the two basins will continue in the years to come to give interesting insights into the issue of international trade in water rights. The reader will be able to access to the relevant news and up-to-date information on the author's website: <www.internationalwaterpolicy.org>.

The ideas developed in this book are first the fruits of my PhD research at the Graduate Institute of International Studies in Geneva from 2003 to 2008. I had the chance to present my work in many places and symposiums such as the ECPR[1] in The Hague, the ISA[2] in Honolulu and San Diego and the World Water Week in Stockholm; these were great opportunities to get feedbacks, to refine and develop my argument. I would like to particularly thank Peter Rogers, Professor of Environmental Engineering at Harvard University whom I met at the 2007 World Water Week in Stockholm for his interests in my work and constructive comments. He persuades me to continue my research efforts and publish my work. I will always feel indebted to him.

I would like also to thank the professors at the Graduate Institute of International Studies for providing me with the tools in law, economics and political science for interdisciplinary work. I am particularly grateful to Professor Urs Luterbacher who supervised my PhD thesis, but also Professors Sylvan, Dupont and Boisson de Chazournes. I am indebted to Professor Dupont and the regretfully departed Doctor Ellen Wiegandt for having offered me the opportunity to have a firsthand experience in international watercourse management by working on a local project on the Rhône River in 2006. I would like also thank Doctor

[1] European Consortium for Political Research

[2] International Studies Association

Marwa Daoudy for her support and for giving me the opportunity to present my results in her class on water conflict and cooperation at the Graduate Institute.

Nell Williams, who is currently a PhD candidate and an assistant in the Political Science department at the Graduate Institute, has done the English revision of this book. For her amazing and professional work, I would like to thank her very much. It goes without saying that I am responsible for all the remaining errors.

Finally, I would like to dedicate this book to my husband Rafael for his loving support, tenderness, patience and faith.

Introduction

Water resource management is one of the most important challenges of the 21st century. Every week new articles and books are published on the subject. The 'water crisis' is being addressed in an increasing number of international symposiums and conferences. Everyone is aware of the need to better manage the world water resources and everyone feels concerned about it. The topic is more often than not argued in highly emotional terms. Today, the debate over water as an economic good or as a human right is an illustration of the polarization and simplification of the issues. This work aims to show that these two concepts are not necessarily mutually exclusive; it all depends on the type of property regimes and institutional frameworks established. To design efficient, equitable and sustainable water policy, there is a need to reconcile the various economic, social and environmental dimensions of water resources.

The principal focus of the present work is on the main controversy today: the economic dimension of water resources. The main objective here is to study the consequences of the recognition of water as an economic good on a particular type of water resources, i.e. international watercourses and aquifers. What will this recognition concretely mean for the management of international watercourses? What types of institutional framework are necessary to accompany such recognition? To what extent is the creation of international water markets a solution to water scarcity?

When I first started studying water issues, it was as a source of conflict in the Middle East region. Discovering more theories of international relations, I entered the fascinating field of hydropolitics (Turton and Henwood 2002). In fact, almost all grand international relations theories can be applied to the study of water relations among nations. Water has become a sort of IR 'laboratory' to

ISBN: 9781843393610. Published by IWA Publishing, London, UK.

develop and test different IR theories and concepts such as hegemonic stability and power relations in river basin (Lowi, 1993), negotiation theory (Faure and Rubin 1993, Daoudy 2009), regime theory (Jägerskog 2001, Jägerskog 2002) or game theory (Güner 1997, Hafterdorn 2000, Luterbacher and Wiegandt 2001).

Some of these studies led me to the unrewarding conclusion that water is both a source of conflict and cooperation. Indeed, water relations including virtual water trade (Allan 1997) create a complex web of geographical, economic and political interdependencies. And as Helen Milner showed for the international system, interdependence does not always lead to harmony or cooperation but rather a mixture of conflict and cooperation (Milner 1993).

With this background in mind, I began to realize there were two types of 'water crises', 'manufactured' crises and 'genuine' crises[1]. The 'water-war' and 'water as oil of the 21st century' discourses have deep resonance in the general public but fail to address the true problems and challenges. The true crisis is that 884 million people do not have access to clean drinking water and that 5 million people die each year of waterborne diseases (WHO/UNICEF 2008). Rivers are polluted and their flows declining, groundwater depleted, ecosystems threatened. These crises are due to inadequate water policies and water management. Transboundary rivers, precisely because they are 'international', are even more concerned by the lack of cooperation and good water governance.

Among the different policy options to improve water management, policies recognizing the economic dimension of water resources are increasingly put forward as an essential component of water reform worldwide. Indeed, in the last few years, various international symposiums such as the 1992 Dublin Conference or the World Water Forums have strongly advocated the recognition of water as an economic good in domestic and international politics. This recognition takes place in a wider context where market-based instruments (MBI) in general are proposed to improve the management of the environment (e.g. EU Commission 2007). The Kyoto Protocol creating international tradable emission permits to reduce in a flexible way the emission of greenhouse gases in the atmosphere is a clear illustration of these innovative policies. MBI have been introduced in many places for different types of resources such as fish, air, and water. In the case of water, basically two kinds of trading systems can be distinguished: pollution permits trading systems and water markets (i.e. trade in water rights). In our discussion over the different policy options recognizing water as an economic

[1] According to the expression used by A. Turton in a keynote lecture at the 2007 World Water Week, Stockholm in the seminar 'Under Cover! Revisited: Groundwater as an Integral Part of Transboundary River Basin Management in Africa', August 2007, Stockholm, Sweden.

good, we will essentially focus on the quantitative aspects of the resource. Therefore we chose to develop 'water markets' options that is roughly quantitative water rights trading systems (although as we will see later on water rights include many more dimensions than just quantity).

It should be clear here that water markets are one policy options among others to recognize water as an economic good and that there exists a wide variety of water markets. The concept of 'economic good' is understood in very different ways among the water community. For some economists, the concept is interpreted in a strict neoclassical sense, with water being treated like any other commodity and allocated through market principles. In this case, economic good means private good (Briscoe 1996). Other economists adopt a more general approach, defining water as an economic good for the simple reason that it is a scarce resource for which there are many competing uses[2] (Perry et al. 1997). According to them, water is clearly an economic good and what matters is the type of economic good it is. In other words, the economic dimension of water resources is separated from the property regime that governs it. In this view, an economic good is not necessarily a private good. In this work, the economic dimension of water resources is understood in a broad sense in such a way that it is compatible with legal, political and environmental approaches.

The concept of property regime also needs to be clarified before going into more details. As we just saw, in the debate over water as an economic good, there is a general confusion between the economic dimension of water resources and the property regime that governs them. Another weakness in this debate is the understanding of property regimes as a black and white issue. This confusion is understandable given the background of the debate.

Indeed, natural resource economics has long been structured along the rift of private versus common property regimes. One of the most influential works in the field is Hardin's *Tragedy of the Commons* (Hardin 1968). In this work, Hardin argued that common good will always be destroyed because the gain that individuals make by over-exploiting it will outweigh the loss they suffer as a result of that over-use (because this loss will be divided among all individuals). He thus demonstrated that the problem of natural resource mismanagement is the by-product of common property. Hardin's article provided an academic rationale for the creation of private property rights systems for natural resources worldwide. But what he described as a commons is in fact a 'free access' resource

[2] Robbins' definition of economics: 'the science which studies human behavior as a relationship between ends and scarce means which have alternative uses' (Robbins 1935). Quoted in Perry et al. 1997.

(i.e. no-property regime) and this confusion over the commons and free access resources remains today.

In this body of literature, other works came to complement this demonstration. Indeed, if one advocates for the creation of private property rights the key question becomes then how to allocate them, how to introduce this new system. To resolve this problem, environmental economists have referred to Coase's work (Coase 1960). Although Coase's demonstration initially concerned harmful effects of production activities, the so-called 'Coase Theorem' has been applied to natural resources management including water resources to argue that from an efficiency point of view, the initial allocation of property rights does not matter under the conditions of no transaction costs and well defined property rights. We will come back later in more detail to the controversial application of Coase Theorem to water resources (part III).

In any case, these two works have durably influenced environmental economics and set the terms of the debate. A common property regime leads to environmental degradation; to ensure an efficient allocation of natural resources, create private property rights and let the market work. A property regime is however never purely common or private. The property regime governing a resource does not depend on the natural characteristics of the resource only. It is also the results of technological development and societal choices.

There exists a wide variety of combinations of property regimes, ranging from fully private to fully common. Property regimes are best represented on a scale combining different degrees of 'privateness' and 'publicness'. Besides, the particular characteristics of each natural resource multiply the possibilities of property regimes. In other words, what in economics is classified under the category of 'common pool resources', can be divided in many subcategories. This means that excludability and rivalry should also be understood in term of degrees.

These remarks hold even truer for water resources. Indeed, water resources are typically multi-use, multi-user resources (Meinzen-Dick and Jackson 1996). The degree of rivalry depends on the type of uses and levels of management. Thus, depending on water availability, irrigation uses can be rival for irrigators of the same districts. But, for example, the rivalry between irrigation uses and hydropower uses is of a different nature because it concerns the time and space of water availability.

In general, the wide variety of water uses are never completely consumptive (industrial uses such as cooling, cleaning processes, irrigation, hydropower, domestic uses, etc.). The omnipresence of return flow therefore complicates the question of rivalry. Besides, water can be treated and retreated. It is not used only once, and so the return flow quality and the level of technological development is important to take into account. Finally, some uses like recreational, aesthetic or environmental uses are usually non-rival because they are non-consumptive uses.

The concept of excludability, like the concept of rivalry, is also best understood in term of degrees. Excludability is determined by the type of uses but also by the technological level of the society. Generally speaking, water resources are difficultly excludable because water is present everywhere under a large variety of forms throughout the hydrological cycle. Water resources are mostly renewable. Besides, the uses of water for the production of particular goods such as river flow regulation, navigation or ecosystem protection are non-excludable goods.

Since water resources are used for a multiplicity of purposes, corresponding to different levels of rivalry and excludability, and because they are both stock and flow resources, they will necessarily be governed by a variety of property regimes including various levels of 'publicness' and 'privateness'.

The various uses of water clearly have an economic dimension. Water is an economic good. The question that needs to be addressed is what type of economic good it is (Perry et al. 1997). The conclusion that can be drawn from the above analysis is that there is no single answer. Water is a multi-property regime resource.

In this work, institutionalist approaches to property regimes are used to complement economic approaches. Thus, property rights are understood as a bundle of rights rather than simply factors of production. They result from a combination of natural and human elements. Among the various rights (and duties) forming a property rights regime, Schlager and Ostrom (1992) identify three distinct classes of rights: operational level rights (access and withdrawal rights), collective-choice level rights (management, exclusion and alienation rights), and constitutional level rights (the right to design collective-choice rules). They explain that in economics the right of alienation (the right to sell or lease management and exclusion rights) is crucial for the efficient use of resources; it forms the separation between truly private and common property rights regime. But here again, there are rules governing the transfer, sale or lease of water rights, and thus the private character of alienation rights depends on the institutional framework. In any case, as suggested by Schlager and Ostrom, it is interesting to complement the definition of property regime by adding decision-making level rules to operational level rules.

Coming back to the object of our study, i.e. transboundary watercourses and aquifers, we now address the question of how to study the property regime of such resources. Suzan Buck (1998) explains that 'just as an individual holds a complex bundle of rights in his or her own society, so does a nation-state exercise a complex bundle of rights in the international arena' (Buck 1998, 27). For us, however, we should not limit the study of property regimes of international watercourses and aquifers to the international level and to state-to-state relationships only. In other words, our approach corresponds to unpack 'the black box of the state' in international relations theory.

Indeed, international water resources involve many different levels of management (international, national, regional, local, etc.) and different decision-making forums at each level of management (e.g. at the national level: agriculture ministry, energy ministry, environment ministry, health ministry, etc). For that matter, the definition of property regime as a 'web of interests' brings interesting insights into the study of property regimes of international water resources. Craig Anthony Arnold (2002) proposes replacing the bundle of rights concept by that of a 'web of interests'. Contrary to the bundle of rights approach, the concept of a 'web of interests' puts at the center of the relationship the characteristics of the object of the property. He defines 'a web as a set of interconnections among persons, groups and entities each with some stake in an identifiable object (tangible or intangible), which is at the centre of the web' (Arnold 2002, 333).

This concept has two merits for our study. Firstly, by placing the object of the property at the centre of the relationship, the concept underlines the particularities of the natural resource and thus enables a greater recognition of the ecological characteristics of each natural resource as inherent limits to its use. Secondly, it recognizes at the same time different types and levels of interest-holders.

We propose applying the web of interests' concept of property to international watercourses and aquifers. For us, water resources have particular natural characteristics (different from other natural resources) that trigger special duties for those responsible for their management. Besides, water property regimes involve many different levels of interest and responsibility. The development of inter-basin transfers, and even international water transfers brings to light the various levels of interest in water management today and raises for example the important question of the right of non-riparian states to an international watercourse. To define the property regime of an international watercourse as a web of interests allows integrating into a single framework the different rules, levels of management and decision-making forums. As illustrated in the figure below, to define the property regime of an international watercourse means to define the architecture of the web - meaning the links between the different levels of management and how to coordinate the decision-making forums-, but also how to modify the architecture of the web (constitutional level rules).

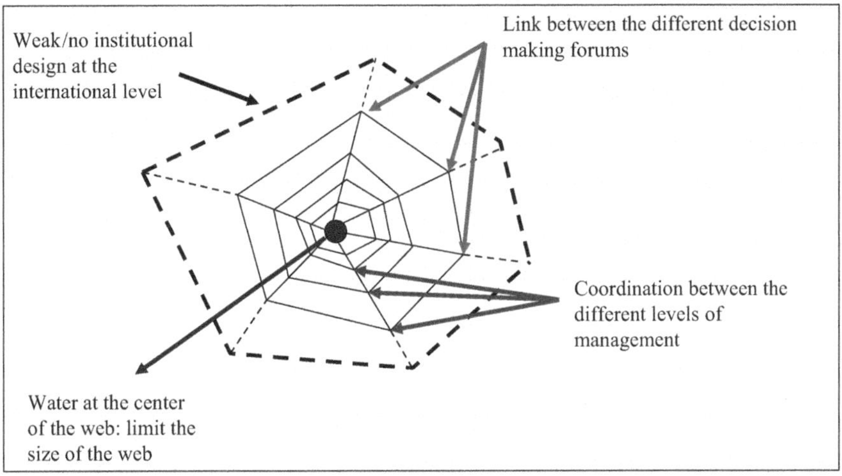

Figure 1 The Property Regime of an International Watercourse as a Web of Interests.

This conceptual framework will be used to address the central question of this work: what are the consequences of the recognition of water as an economic good in domestic laws on international watercourses management?

It is argued that, as these various policies move toward fuller implementation, they will increasingly impact the way international watercourses are managed. As an example, the clarification of water rights and the creation of an administrative system of water rights at the domestic level mean that sooner or later water rights over international watercourses will have to be clarified as well. The same is true for pricing policies. These policies could lead to inefficient and inequitable results for international watercourses if they are not accompanied by the implementation of a strong institutional framework and a clear property regime at the international level.

Moreover, the lack of a clear property regime on international watercourses is put under light by the increasing application of international trade rules to water resources. Indeed, referring to Chichilniski's analysis (Chichilnisky 1994), it is argued that international trade rules are applied by default precisely because there is often no clear and enforced property regime on international watercourses. International water transfers, especially those involving international watercourses such as transfers from the Great Lakes have recently been the illustration of the tension between international water law principles and international trade rules.

But more than a clash between international water law and international trade rules, what one observes is a disagreement over the actors, the principles and the process of decision-making concerning the management of the shared resources. In other words, the tensions are occurring over the constitutional level, i.e. over the rules to design decision-making rules.

In this context, the introduction of international water markets will not help resolve international water disputes. In most economists' proposals on international water markets, the institutional aspects are neglected. What would it really require to introduce international water markets along international watercourses?

METHODS

This work was conducted as policy-oriented research. The general topic of this book is water management or, in more fashionable parlance, water 'governance'. The central objects of the study are institutional designs, decision-making procedures and water rights. These elements are all included in our key concept of 'web of interests' that we defined above and that we will use to study the property regimes of water resources. Our policy recommendations are the results of our case studies. The two cases of the Colorado River Basin and the Murray-Darling Basin were selected because they are the only two examples of *inter-state* water rights trading systems in the real world. Because these cases are different in many respects, the inductive method and process-tracing technique were preferred over other methods (Bennett and George 2001). To conduct this study we used different types of documents such as national water legislation (FAOLex), official reports on water reforms, policy papers and technical papers by the World Bank, international and national court's cases but also other documents issued by river basin commissions and water agencies (meetings' minutes, annual reports, agreements). Most of our primary sources are available on the internet. The documents of the negotiations of the drought plan on the Colorado River Basin including draft agreements, basin states' letters, and public comments are all available on the Bureau of Reclamation's website. The Murray-Darling Basin Commission/Authority's and the Department of the Australian Prime Minister's websites release also many documents concerning the water reform, the negotiations and the implementation of the interstate trading scheme. Finally, we had direct email contact with some of the practitioners, for example, managers of the Southern Nevada Water Agency.

The present book is structured in three parts. In the first part, we study the significance of water as an economic good and how it is been implemented in various countries. We will present current trends in domestic water reforms while showing the implications for international watercourse management. An important piece of this first part will be devoted to the study of one particular type of policy option, i.e. water markets. Water markets receive a substantial amount of attention today; they are increasingly presented as the model for water reform worldwide by some influential economists.

The second part of this work looks at the relationship between trade law and international water law. The objective of this part is to show that trade rules are increasingly applied to international water resources – by default - because there is a lack of clear property regime rules governing this type of water resources.

The third part is devoted to the analysis of market-based solutions in international river basins. The theoretical and practical weaknesses of various projects of international water markets are presented. Then, in order to have a better understanding of the true challenges raised by the introduction of water rights trading schemes along international watercourses we study two current initiatives in two large river basins. These two case-studies are the Colorado River in Western USA and the Murray-Darling River system in Australia. These two cases are not international but we believe that these two federal cases can bring interesting insight into the discussion of market-based mechanisms along international watercourses.

PART I

Water as an Economic Good in Domestic Water Reforms

Over the last decade, many countries worldwide have adopted new water legislation revealing a global evolution in water management, illustrated mainly by the formalization of water rights and entitlements (Spain in 1985 and 1999, Mexico in 1992, Madagascar in 1998, South Africa in 1998, China in 1998 and 2002, Vietnam in 1998, Kenya in 2002, Brazil in 1997 and 2000, Australia in 2004, Indonesia in 2004, etc.). The formalization of water rights, the introduction of new pricing policies or the establishment of tradable water right systems are all different illustrations of the worldwide trend towards the recognition of water as an economic good.

Although water has been recognized as an economic good for a long time in some parts of the world, it is a recent development in most countries. In order to stress the significance of recent water reforms, the historical evolution of water law is retraced. To study water policies at the domestic level is not a luxury but rather an essential step before highlighting the consequences of these policies on international watercourses management.

In this global movement of water law modernization, one can distinguish a dominant trend toward the creation of tradable water rights (Chile, Australia, USA,

ISBN: 9781843393610. Published by IWA Publishing, London, UK.

Spain, Mexico). Tradable water rights are indeed perceived as promising water management tools by leading institutions such as the World Bank (World Bank 1993). However, the concept of water market includes many different realities. Our analysis of water markets will identify to what extent and under which conditions they can improve water resource management.

But before turning to these issues, we will present the different conceptual approaches of water as an economic good.

Chapter 1

The International Recognition of Water as an Economic Good

CONTROVERSIES OVER WATER AS AN ECONOMIC GOOD

The explicit recognition of water as an economic good at the international level is a recent development. It is usually acknowledged that water was first overtly recognized as an economic good at the 1992 Dublin Conference on Water and the Environment (ICWE 1992). Organized by the World Meteorological Organization (WMO), the conference gathered five hundred participants, including government-designated experts and representatives of eighty international, intergovernmental and non-governmental organizations in January 1992 in Dublin to prepare recommendations for the Earth Summit held few months later in Rio de Janeiro. At the end of the preparatory session, the conference participants issued a set of recommendations for action at local, national and international levels based on the following four guiding principles:

> 'Freshwater is a finite and vulnerable resource, essential to sustain life, development and the environment; Water development and management should be based on a

ISBN: 9781843393610. Published by IWA Publishing, London, UK.

participatory approach, involving users, planners and policy-makers at all levels; Women play a central part in the provision, management, and safeguarding of water; Water has an economic value in all its competing uses, and should be recognized as an economic good' (ICWE 1992).

The United Nations Conference on Environment and Development held in September 1992 in Rio recognized also water as an economic good:

'Integrated water resources management is based on the perception of water as an integral part of the ecosystem, a natural resource, and a social and economic good'(UN Agenda 21, Chapter 18.8).

In the follow-up process of the Earth Summit, two umbrella organizations were created outside the UN framework in 1996. The Global Water Partnership was created by the World Bank, the United Nations Development Program (UNDP) and the Swedish International Development Agency (SIDA). The World Water Forum was created by senior members of the International Water Resources Association (IWRA). Both institutions are composed of governments, multilateral banks, UN agencies, professional associations, research organisations, private companies and NGOs. They both organize international conferences and are committed to providing tools for discussion and implementation of the Dublin-Rio principles.

The recognition of water as an economic good has raised significant controversies amongst the civil society organizations who organized various alternative forums, especially during the 'official' World Water Forums (1997 Marrakech, 2000 The Hague, 2003 Kyoto, 2006 Mexico, and 2009 Istanbul). The NGO statement at The Hague alternative forum mentions for example that: 'Access to basic water and sanitation are universal rights and cannot therefore be negotiated as commodities.' (These reactions are essentially directed against the global trend toward water privatization that is the privatization of water services. Although both issues are closely linked, it is essential to distinguish privatization of water services from privatization of the resource itself. The subject of this research focuses on management of the resource itself).

For most water activists, the recognition of water as an economic good - because it is understood as commodification of the resource - is not compatible with the recognition of water as a human right[1].

[1] The global water civil society is divided on the content of the human right to water. For some (e.g. the Council of Canadians, etc.) the implementation of such a right necessarily implies to strictly oppose any recognition of water as an economic good. They are clearly against the 1992 Dublin statement mentioning that 'water has an economic value in all its competing uses and should be recognized as an economic good' (International Conference on Water and the Environment 1992). Because the

In November 2002, the United Nations Committee on Economic, Social and Cultural Rights, which monitors the International Covenant on Economic, Social and Cultural Rights declared explicitly access to water a human right and recognized water as 'a public good fundamental for life and health' (General Comment n° 15 on articles 11 and 12 of the Covenant). In this declaration, state parties are recalled of their responsibility to implement the right to water like any other human right recognized in the covenant.

The recognition of water access as a basic human right is essential to correcting economic approaches, but these two policies are not contradictory. Peter Gleick notes indeed that: 'a broad economic approach to water management does not inevitably lead to the management of water as if it were a commodity in all its aspects' (Gleick *et al.* 2002, 37). Perry, Rock and Seckler clearly explain that 'the question is not whether water is an economic good or not but what kind of economic good it is'. It is a societal choice to decide if it is a private or a public good - or a combination of both. To recognize water as a public good implies that the true 'economic' value of water is different from its financial value - that is the value based on market prices only (Perry *et al.* 1997). Then the question of how to finance the access to water as a basic human right remains.

Various national reforms attempt to reconcile notions of water as an economic good and water as a human right. One can distinguish two main orientations in achieving this goal. As is the case of most developed countries that tried to favor the human right to water, France adopted legislation requiring that 'Each individual, for alimentation and hygiene, has the right to access drinking water in economic conditions affordable by all' (Article 1 of the 2006 water law). The issue here is to define what an affordable price for drinking water is[2]. There is no general rule to determine it and the concept of 'water affordability' can vary significantly depending upon the level of development of a given country. In some countries, public authorities deliver targeted financial help to help poor households pay their water bills. In others, poor households have special water tariffs.

South Africa - which was one of the first countries in the world to explicitly recognize in its constitution the right to access water - adopted a more radical approach to the human right to water. The 1996 South African Constitution

convention proposed by Green Cross International recognizes water as both a human right and economic good, this former group of NGOs refuses to support GCI's project and sign on its petition for the convention.

[2] Henri Smets (Smets 2008) proposes to determine 'economic affordability' by an index comparing the water and sanitation bill of a household to its income.

recognizes 'the right to have access to sufficient water' (Bill of Rights, Constitution of South Africa, Section 27(1)b). In order to implement this right, in 2001 President Mbeki launched the 'free basic water policy'. It required that progressively 6,000 litres of clean water be provided free to every household each month and available within 200 meters of the household. This amount of basic water was determined based on the World Health Organization's recommendation that each person needs between 20 to 50 litres of clean water per day and estimating that on average eight persons live in any given household (25L/pers./day). However, in many cases, this basic amount of free water is insufficient and falls below the WHO's limit as most households have more than eight members. In fact, it is up to the local governments to implement this policy according to local conditions as they are the ones responsible for providing clean water services. Many local governments, however, lack funding – even with cross-subsidization – to implement this free basic water policy and develop the necessary infrastructure. Implementation was also rendered difficult because it was associated with a full cost recovery policy and the 'water pays for water' principle. As a result, the price of water to be paid for the water used above the free amount is too high for many households. Thus, even if poor people are provided a certain amount of clean water for free, they have difficulties paying their water bills. This problem has been accompanied by the controversy over pre-paid meters in Johannesburg which cut off water supply in cases where no payment was made.

The South African experience shows that the implementation of the human right to water while simultaneously recognizing water as an economic good is a challenge but not impossible. According to the DWAF, Department of Water Affairs and Forestry, free basic water was provided to 86% of the total population as of January 2010 (DWAF 2010).

WATER RESOURCE ECONOMICS

Today, we observe a paradigm shift in water resource management. Demand management policies are progressively taking over supply enhancement policies. When the supply of water resources did not meet demand, technological and engineering solutions were found to enhance water supply and thus meet increasing water demand. For decades, the logic was to build more and larger dams to store more water, to drill wells, to construct pipelines to transfer water from water rich regions toward water scarce regions. However, these projects are becoming more expensive, technologically challenging and environmentally damaging with the passing of time. Thus, alternative policies focusing on demand management have started to replace the former logic of supply enhancement. Whereas supply enhancement policies require good engineers, demand management policies require good economists (and good engineers!). The application of economics to a resource such as water is very challenging. Economists are still trying to clarify to what extent the economic discipline can contribute to good water management and often disagree on this very subject.

A first difficulty for water resource economists is to measure how much water is available for the society, or, in other terms, to determine water supply. Water is everywhere and takes very different forms according to the location, climate and time of the year. The water cycle implies many complex relationships among the different forms of water, the various water uses and the larger environment. Because water is everywhere, macroeconomic policy influences water uses in many ways. Even if the policies do not directly deal with the water sector, many of them impact it anyway. This is the case, for example, of energy policy, agriculture policy, tourism policy or even trade and exchange rate policy (e.g. when designed to promote water-consuming industries such as the chip industry). Economists have to keep these things in mind when designing specific economic instruments for the water sector. They also clearly need to work in close collaboration with other disciplines (particularly natural science) to determine the sustainable limit of water use.

Another difficulty for economists is to measure the overall water demand. Water is used for different types of demand: some uses deal with the stock aspect of the resource, while other uses require the flow aspect of the resource. Each use has a different level of 'consumption' of water, and water can be reused.

Finally, another big challenge for economists is how to determine and compare the values of water resources for different users given that they value differently the various types of water resources (e.g. water with different levels of treatment).

Each economist has their own understanding of these problems and proposes different solutions. Before going into further detail, we would like to present some basic elements.

There exists an important difference between water as a resource in nature, bulk or wholesale water and water that arrives in the tap (retail water). Some water uses do not require any transformation of the resource; this is the case, for example, of rain-fed agriculture or recreational activities in natural lakes and streams. Bulk or wholesale water corresponds to water with some degree of transformation. In the case of surface water, bulk water management means the construction/maintenance of reservoirs or artificial lakes, pipelines and other infrastructure to carry water to bulk users: it can be, for example, a municipality, an industry, a retail operator or an irrigation district. In the case of groundwater, bulk water costs include pumping costs. Depending on the regulation of each country, this first transformation is operated either by the state, a basin agency or a bulk water company. In some parts of the world, an increasingly important activity for bulk water providers is desalination.

Retail water, which is water delivered to the final consumer, requires further degrees of transformation and treatment. Retail water production and distribution activities are usually associated with wastewater collection and treatment. These services are undertaken by various types of utilities including private, public and combination of both.

The Cost of Water

Economists usually agree on the supply side of the equation: i.e. the cost of providing the resource. For them, the cost of providing the resource should correspond at least to the production costs (recovery costs). Production costs should take into account construction and maintenance of infrastructures such as conveyance facilities (canals, pipelines, rivers) and storage facilities (reservoirs, lakes). Production costs should also include administrative services such as planning, management, monitoring and billing. Thus, even bulk or wholesale water has a cost. Then, the cost increases with the level of transformation of the resource before it reaches the end users. For example, the production costs of retail water to urban households should include – in addition to the above costs-treatment costs (depending on the quality of bulk water), storage and security costs (to assure reliable supply even in case of shortages) and costs to pressurize water (energy costs, tower water).

All these costs are components of the total production costs. What is even more important for economists is to calculate the marginal cost of the resource, i.e. the cost of producing one more unit of water. Indeed, the costs of production usually depend on the quantity of the products produced. In the case of water resources, this clearly complicates the task of the economists. Griffin (2006) stresses the particular case of urban retail water distribution. He notes

> 'In water supply circumstances, there can be major costs that are largely unrelated to water deliveries. An important example is the urban distributional networks that connect all residences to a central water supply. This distribution system is costly regardless of how much water is passing through it' (Griffin 2006, 16).

Although economists generally agree on the importance of recognizing all these elements forming the production costs of bulk and retail water, their measurements remain a challenging task.

Controversies and disagreements arise with regard to the opportunity cost of water resources. Should opportunity cost be added to the production costs of water resources? And more precisely, how should it be measured? Here is the central point of the controversy over water as an economic good. Basically, mainstream economists and other economists split on this particular point. Opportunity cost is not a neutral concept; it triggers water allocation issues and deals with the question of the 'value' of water.

Opportunity Cost, the Scarcity Value of Water Resource and Water Markets

Mainstream economists usually underline the importance of the recognition of water's opportunity cost for efficient water use and allocation. Briscoe (1996) defines opportunity cost as 'the cost incurred when one user uses water and affects the use of the resource by another user' (Briscoe 1996, 8). Another common definition of opportunity cost is 'the value of water in its highest value alternative use,' which is supposed to reveal the scarcity value of water resources.

Briscoe explains that

> 'It is obvious that measuring the opportunity cost of water is a difficult task. [. . .] What can be said with certainty is that: Opportunity costs are related to value in a non-transitive way. That is, if a city and an irrigation district lie on opposite banks of a stream, the opportunity costs imposed by abstraction by the high-value user (the city) will be much lower than the opportunity costs imposed by abstraction by the low-value user (the irrigation district); Opportunity costs increase substantially as the water in a basin becomes more 'densely used'; The existence and imposition of opportunity costs can give rise to conflicts among users, unless there are institutional mechanisms for recognizing these costs and for ensuring that these are taken into account by users' (Ibid., 8-9).

Asad *et al.* (1997) also recognize that it is difficult to measure opportunity costs and to integrate them into water pricing. They explain that reformers will face enormous opposition if they want to integrate opportunity cost in their pricing policies. Indeed,

> '[. . .] socioeconomic realities in most countries, particularly developing countries, would render any implementation of opportunity cost pricing highly impractical' - at least for two reasons: 'because the information requirements are onerous (opportunity costs vary dramatically by place and season) and 'because levying such charges would (usually correctly) be perceived as expropriation by those who currently use the water' (Asad *et al.* 1997, 9).

They therefore conclude, like John Briscoe and other mainstream economists, that the best solution to integrate opportunity costs, i.e. the scarcity value of water, is to create the conditions for water markets to emerge. For John Briscoe, giving the example of the Chilean water market, 'it is evident that no administrative mechanism can mimic water markets in transmitting information on opportunity costs in such a flexible and specific way' (Briscoe, 1996). He also notes that 'this is the genius of the water market approach – it ensures that the user will in fact face the appropriate economic incentives, but de-links these incentives from the tariff (which is set on 'common-sense' ground)' (Briscoe 1996, 18).

The superiority of water market as reallocation mechanisms is showed in another World Bank study comparing four types of water allocation mechanisms (Dinar, Rosegrant and Meinzen-Dick 1997). Dinar *et al.* analyse the advantages and disadvantages of each of four different water management policies: marginal cost pricing, administrative/governmental allocation, water markets and user-based allocation. Administrative government allocation is described as the worst solution: they explain 'supplying water to deficient areas leads to expensive, publicly financed water projects which preclude any need to purchase water rights based upon the scarcity of the resource. Public allocation mechanisms often lead to waste and mis-allocation'. Asad *et al.*, analysing these results, declare: 'At least one message is clear: public water allocation is the least effective and efficient mechanism' (Asad *et al.* 1997, 10).

Another mode of allocation studied by Dinar *et al.* is marginal cost pricing, i.e the price of water to equal the marginal cost of supplying the last unit of water. The idea for those supporting this policy is that if prices are right, allocation will be optimal. This mechanism has two main disadvantages according to Dinar *et al.*: it is very difficult to calculate the true marginal cost and it does not include the scarcity value of the resource.

According to them, user-based allocation mechanisms have the good characteristic of been flexible and adaptable to local conditions, they are however often limited to one particular sector and do not offer solution for intersectoral allocation.

Finally, they present water markets as 'not a panacea' but nevertheless the best solution for two reasons. As discussed above, water markets are the best method to integrate the full economic cost of water resources including opportunity cost. Besides, the creation of water markets is perceived as the best and softest means to reallocate water resources across sectors. Dinar *et al.* indeed mention 'moving toward tradable property rights in water may ease the process of intersectoral reallocation by compensating the 'losers' and creating incentives for efficient water use in all sectors'(Dinar, Rosegrant and Meinzen-Dick 1997).

Alternative Understanding of Opportunity Cost, Scarcity Value and Efficiency

For some economists, the recognition of opportunity costs does not necessarily imply the creation of water markets. Griffin is more careful about the notions of opportunity cost and scarcity value of water resources. In order to determine the scarcity and availability of water resources, it is necessary to measure water demand. He notes however that the measurement of the aggregate demand of water resources 'is trickier for water than it is for most commodities', therefore water scarcity is also difficult to measure. He gives three reasons: 'differential processing' (i.e. different degree of transformation of water for different uses), 'reuse' and 'nonrivalness'(Griffin 2006, 29).

In addition, for him, the question of 'which sector values water more highly' is usually wrongly answered.

> 'A common misstep in water resource discussion is to casually say that urban users value water more highly than irrigators. This claim might be true in some sense, but one must be clear about what is being discussed – retail or natural water. [. . .] To infer their implied natural water values correctly and thereby enable an accurate comparison, one should subtract all the embedded processing costs from the retail water prices. Often, the correct procedure will identify a zero natural water value for both sectors because there are no price-embedded costs associated with natural water, especially in water-rich regions, but often in arid regions too. That natural water is unvalued is a distressing finding, but it is an honest one, and it clearly underscores one of the major policy flaws in water resource management' (Griffin 2006, 38).

Griffin's analysis that the value of natural water is not really significant might be questioned especially in arid regions. He is right however to stress the problem of the comparison of the value of water across different sectors using very different levels of water treatment and quantities of water such as the agricultural and urban sectors. The two sectors do not value the same 'product'. And because the two sectors are generally competing for very different quantities of water, the opportunity cost will rapidly drop to zero in the case of a water transfer from the agricultural to the urban sector. Once the demand of urban consumers is satisfied - and this does not require huge quantities of water-, their willingness to buy one more unit of water at a price higher than agricultural users will sharply decrease.

In order to avoid 'apples-to-oranges issues' in the comparison of water values across different sectors, Griffin proposes to study marginal net benefit functions. 'By definition, marginal net benefits equal marginal benefits (demand) minus marginal costs (supply, almost).' Griffin explains that marginal benefits and demand are two terms that are usually used interchangeably and that we should be careful about the terminology. Indeed 'sometimes the word demand is thought to suggest the presence of a true marketplace, which is often false in the case of water because of the absence of either competition for retail water or private property rights to natural resource. In such situations, marginal benefit can be a preferred term' (Griffin 2006, 34).

In any case, for him the calculation of the net marginal benefit function should be an important tool for decision makers.

> 'If a group of agents is taking water from a watercourse and the subtracted marginal cost function is derived from a cost function that includes all operations performed in the withdrawal, processing, and delivery of this water, then the marginal net benefit of retail water is the agent's marginal benefit of natural water in the watercourse. This is crucial information. To say it another way, the net marginal cost of retail water is the marginal benefit of natural water before it is removed from the watercourse' (Ibid., 36).

By comparing the net marginal benefits functions of different users, decision makers can identify the optimal allocation of limited water supplies across different sectors.

Entering the field of optimal allocation, Griffin is very careful. To define the optimal allocation of water resources usually means allocating water resources to their most efficient uses. The search for economic efficiency in a society should be one objective among others. Therefore, he proposes a broad approach to economic efficiency. He distinguishes between two types of efficiency: 'aggregate' and 'neutral' efficiency. The aggregate efficiency usually determines a single efficient allocation only whereas the neutral efficiency offers a range of allocations that are efficient.

> 'Aggregate efficiency weights by income and commodity prices. Neutral efficiency does not weight at all and therefore indicates an array of efficient selections instead of one. Because there are good reasons for income/price weighting in a well-functioning market economy, aggregate efficiency is sensible and has the advantage of resolving a single efficient allocation. But circumstances can arise in which the implicit weights embodied in aggregate efficiency are disagreeable, or are possibly so. In such cases, neutral efficiency can be an illuminating social objective for examining particular water resource issues' (Ibid., 51).

Thus, for Griffin, economic analysis can bring interesting tools for decision makers. However, for him, we also need to acknowledge some weaknesses in the economic discipline, especially when aggregate efficiency is the unique goal.

Griffin's considerations on economic efficiency are linked to the more general discussion on the role of water pricing policies. There is an important difference between pricing policies aimed at water reallocation and those aimed at water conservation. In addition, for many non-mainstream economists pricing policies are only one tool among others to control water demand. Besides water pricing, water management policies include a large variety of tools such as quotas, ration plans in case of drought, incentives, grant or subsidies for water-conservation devices (including as varied things as low flow toilet and showerheads to canal lining), licences, tax or penalties, as well as education and training about conservation options, and development of wastewater treatment (Savenije and Van der Zaag 2001, Griffin 2006). These different understandings of water as an economic good - including among economists- are well illustrated by the internal debates about water policy at the World Bank.

In its 1993 official policy paper on water policy, the World Bank argues that underpricing and misallocation of water resources are the major problems of current water management. Endorsing studies showing that demand is responsive to water prices - i.e. demand is reduced when prices increase - World Bank's economists propose to develop better pricing techniques expecting this would automatically lead to the reallocation of water resources from low-value uses toward higher-value uses. The policy paper mentions further that more accurate water pricing should include opportunity costs of water. However the paper does not clearly present what this would imply. Overall, the paper remains vague about the different aspects of water pricing revealing the state of the internal debates at the World Bank.

Therefore, the World Bank's official position regarding opportunity costs and water market is more careful than the one presented in some of the technical papers referred to earlier (Briscoe 1996, Asad and al. 1997, Dinar *et al.* 1997). In its 1993 policy paper, the World Bank clearly asserts that its priority is first to make sure that water pricing systems include the full use cost, that is at least the cost of recovery (what we called 'production cost'). The Bank stresses also the importance of including opportunity cost in water pricing but acknowledges the numerous social, technical and political obstacles to achieve this goal (World Bank 1993, 10). It then stresses the need for further research to develop balanced institutional frameworks to 'both harness the efficiency of market forces and strengthen the capacity of governments to carry out their essential roles' (World Bank 1993, 10). Disagreements at the World Bank deal essentially with the critical question of the exact level of governmental intervention in water markets and the exact 'breadth' of market institutions.

Despite some precaution and prudence in World Bank policy official position, one notes a leading trend in World Bank policy toward the formalization of water rights and the introduction of trading systems. Such policy orientation was clearly stated in a 2005 World Bank Report on 'India's Water Economy'. However, in practice, as one will show later, 'water markets' encompass many different realities. Beyond the current debates inside and outside the World Bank over water markets and tradable water rights systems, one notes a worldwide trend in domestic water law toward the introduction of formal water rights systems. Although these systems of formal water rights may not be aimed at the development of water markets, it is important to underline that they represent a preliminary and necessary step in the creation of these markets.

Chapter 2

Evolution of Water Rights and Domestic Water Reforms

HISTORICAL PERSPECTIVE ON WATER RIGHTS

To underline the importance of today's water rights reform, it is essential to put it in a broader historical perspective. Considering the huge differences in geographical, institutional, cultural, social and economic context, it will be impossible to analyse water rights evolution in great detail. It seems however important to stress some general basic trends and thus better understand the current shift in water management at the domestic level.

In water management, like in other fields, important technological and economic developments pushed for the introduction of new rules and regulations. The industrial revolution has led to the development of new rules governing the utilization of watercourses. In the development of modern water law, the Code Napoleon or French Civil Code is widely recognized among lawyers as the first coherent statutory and codified body of rules regulating the uses of rivers, streams, lakes and groundwater. It had enormous influence in Europe, the United States and England, and thus part of the Commonwealth. Wiel demonstrates how

 International Trade in Water Rights: The Next Step. By Aline Baillat.
ISBN: 9781843393610. Published by IWA Publishing, London, UK.

the modern doctrine of riparianism was first codified in the French Civil Code, then reached the American Continent, and from there traveled to English Courts (Wiel 1918).

Roman Law

Roman law has been subject to very different interpretations and its coherence has sometimes been put under question (Wiel 1918, 255). It remains however an important body of rules on which rests most of modern law tradition. Roman law distinguishes between public and private rivers. All perennial rivers were considered *Res Publica*, things owned by the public and administered by the state. Torrential streams or temporal streams were considered private property by the persons whose land they crossed. However, under the general classification of things of the Institutes of Justinian (333–334 A.D) the flowing water itself was said to be *Res Communis*: 'by natural law these things are common to all : air, running water, the sea, and as a consequence, the shores of the sea'. By its very nature, water was considered as not being subject of ownership and therefore was neither state property nor private property. Therefore water rights could only be usufructuary rights. Water rights were not necessarily linked to an individual property of riparian land. Water rights could be obtained by 'prescription' as long as the use of water did not change the flow of the river from that of the 'previous summer'.

The two major characteristics of Roman water law is the classification of flowing water within rivers as *Res Communis* and the distinction between public and private rivers. These two major elements are still important in water law today. Although water can now be more easily appropriated because of new techniques such as large-scale diversions, storage dams and underground reservoirs of river water, etc., water rights are usually usufruct rights. All systems of water law have – in theory - adopted this fundamental principle. The distinction between public and private water was taken up in civil water law but today all waters are increasingly declared to be in the public domain.

English Common Law

It is not our intention to retrace in details the evolution of water law in common law[1]. Anthony Scott, Georgina Coustalin (1995) show the complex 'twist and

[1] For more details on the subject, see: Joshua Getzler 2004. *A History of Water Rights at Common Law*, Oxford: Oxford Press.

turn in water law,' illustrating thus how water rights were successively and alternatively based on land ownership or on actual uses. Few cases were recorded in medieval England, but most disputes did not arise over competing uses but rather over the best *sites* for mills. In medieval common law, rivers and streams were conceptualized as a feature of land. Water rights were thus purely based on land ownership and the owner had full right over its water. Scott and Coustalin mention the existence of different phases in the medieval period. Progressively the rights of the owner of land were altered and rights of non-riparian increased. For example, landowners could not preserve their rights for future uses; their rights were tight to their existing uses and had to have suffered damages to an existing use to bring action.

Scott and Coustalin identify a second period of common water law in England, that of prior use from approximately 1600 to 1850, marked by rapid economic development and population growth. Disputes then arose mainly over diversions and mills competing for the flow. Judge Blackstone's writings were essential in the development of this new doctrine. Basing his argumentation on the Roman concept of '*Res Communis* ', he developed a natural law version of the 'prior use right'. Sir William Blackstone explained that water was not capable of private property but of temporary appropriation only. Every man, riparian land owner or not, had a right to enjoy the watercourse without disturbance of existing uses. In case of drought, senior water rights had priority over junior rights. The right to use water stopped when water was not used anymore.

A third period is marked by the progressive development of the different versions of the riparian rights doctrine. Scott and Coustalin explain that the prior use doctrine was first altered by the natural flow theory. Although not applicable to water abstraction but to the water pollution problem alone, the natural flow theory served as a basis for the development of the reasonable use doctrine. Indeed, the natural flow doctrine, in which riparian landowners have equal rights to the natural flow of the river, marked the return to land-based water rights.

A more sophisticated form of land-based water rights was then developed. The introduction of the reasonable use doctrine is dated to 1851 in *Embrey v. Owen*. Wiel explains that this judgment found its legal basis in American judgments which are supposed to have been influenced by the redaction of the French Civil Code in 1804. In any case, the reasonable use doctrine enabled the protection of the industrial development of rivers to a certain reasonable extent. The characteristics of this doctrine are as follows: the riparian landowners had the right to use the river as long as they did so in a reasonable manner. Riparians also had the right to continue to receive the flow even though they were not using or diverting it. This doctrine distinguished ordinary uses from extraordinary uses. For extraordinary uses (i.e. commercial, industrial) riparians could not take action unless the uses were unreasonable and creating excessive damages. Thus, concerning these extraordinary uses, Scott and Coustalin mention that some kind of seniority rule was maintained. Concerning ordinary uses (i.e. domestic), the rule meant that, in times of shortage, the ordinary user located upstream might take all the flow without having to share with other riparians. Finally Scott and Coustalin explain that certain uses of water, such as polluting uses, those uses which carried water off riparian lands and out of the river basin and wasteful uses were deemed unreasonable whether or not they caused actual damage. In any case, the common law version of the riparian doctrine was enforced only by riparians and not by the state. In practice, any use of the water could be made until the user was sued.

French Civil Law

There are few writings and records of the evolution of French water law before the introduction of the Civil Code in 1804. The French Civil Code or Code Napoléon provided clear rules for the uses of rivers, streams, lakes and underground waters[2] and strongly influenced many water law systems in Europe and other parts of the world. It maintains the distinction between public and private rivers found in Roman Law. Article 538 mentions that navigable and floatable rivers are dependent of the public domain. These waters are administered by the government or departmental authorities who may grant riparian owners the right to take water and tolerate the public use of these waters for certain domestic purposes[3].

Articles 644 and 645 regulate the use of waters outside the public domain that is rivulets, temporary streams and other non navigable and non floatable rivers[4]. The French Civil Code establishes land-based water rights for these rivers. Article 645 mentions:

---

[2] Underground waters are governed by article 552 of the French Civil Code. The proprietor of the surface is proprietor of what is beneath.

[3] Labori, Répertoire du Droit Français, quoted in Wiel 1919, 152.

[4] French Civil Code 1804- Article 644 : 'celui dont la propriété borde une eau courante, autre que celle qui est déclarée dépendance du domaine public par l'article 538, au titre de la distinction des biens, peut s'en servir à son passage pour l'irrigation des propriétés. Celui dont cette eau traverse l'héritage peut même en user dans l'intervalle qu'elle y parcourt, mais à la charge de la rendre, à la sortie de ses fonds, à son cours ordinaire.'

He whose property borders on a running water, other than that which is declared a dependency on the public domain by article 538, under the title of the Classification of Things, may serve himself from it in its passage for the watering of his property. He whose estate such water crosses is at liberty to use it within the space which it crosses, but on condition of restoring it, at its departure from his land, to its ordinary course.'

> 'If a contest has arisen between the proprietors to whom these waters are useful, the courts, in deciding should make the interest of agriculture conform with respect to property of land; and, in all cases, the particular and local rules upon the flow and use of waters should be observed[5]'(Wiel 1918, 265).

The French riparian law doctrine provides a perfect equality of right between riparian owners. Their rights do not depend on existing uses but on ownership of riparian lands only. Contrary to prior use systems, rights are not absolute but relative to other riparian rights. Article 645, by giving power to the courts to apportion water among riparians on a case by case basis, opens the way to a flexible rule of reasonable use doctrine. Moreover, French water law is not contained in the Civil Code alone but in legal commentaries as well. Commenting on article 645, Laurent mentions :

> 'Ils ont tous un titre égal; on serait donc tenté de décider que leur droit doit être réglé d'après la stricte loi de l'égalité, c'est à dire, d'après l'étendue de leurs héritages. Mais il y a encore d'autres éléments dont il faut tenir compte, le mode de culture, la nature du sol, le genre d'exploitation. Il est impossible d'établir une règle absolue et une égalité mathématique : voilà pourquoi la loi s'en est rapportée à la sagesse des tribunaux'[6] (Laurent, 1873, Principes du Droit Civil, Tome 7, p326, quoted in Wiel 1918, 259).

Reasonable and Equitable Use Doctrine

The term 'reasonable' appeared first in an American court case, *Tyler v. Wilkinson,* rendered in 1827 by Justice Story in the US circuit court for the District of Rhode Island[7]. In his opinion, Justice Story recognized the civil riparian law : 'The natural stream, existing by the bounty of Providence for the benefit of the land through which it flows, is an incident annexed by operation of law to the land itself'[. . .] 'There may be, and there must be allowed to all, of that which is common, a *reasonable* use'[. . .] 'It is not like the case of mere occupancy, where the first

---

[5] French Civil Code 1804 Article 645 : 'S'il s'élève une contestation entre les propriétaires auxquels ces eaux peuvent être utiles, les tribunaux, en se prononçant, doivent concilier l'intérêt de l'agriculture avec le respect dû à la propriété ; et, dans tous les cas, les règlements particuliers et locaux sur le cours et l'usage des eaux doivent être observés'.

[6] They all have an equal entitlement; one could thus be tempted to decide that their right be governed by a strict equality rule, that is, according to the area of their heritages. Other elements should however be taken into account, the crops cultivated, the nature of the soil, the type of farm. It is impossible to establish an absolute rule and a mathematical equality: this is the reason why the law referred to the wisdom of the courts.

[7] Tyler v. Wilkinson, 24 Fed. Cas. 472 (N° 14,312) (CCDRI 1827).

occupant takes by force of his priority of occupancy'. Thus, Justice Story rejected prior use doctrine, holding that each riparian owner has a right to a reasonable use of the water. The reasonable use version of riparianism is to be distinguished from the English natural flow version. In his famous *Commentaries on American Law*, published in 1928, Kent cited *Tyler v. Wilkinson* in his discussion of water rights. In fact, he found the origins of the 'just and equitable principle' in the French Civil Code, the works of French lawyers like Touiller, Pothier and Merlin and Roman jurists[8]. He quotes the Roman maxim: *'sic enim debere quem meliorem agrum suum facere, ne vicini deteriorem faciat'*[9] to explain clearly that riparian rights allow consumptive use as long as it is 'reasonable'. In sharp contrast with Civil Law, however, the Common Law of riparianism does not distinguish between public and private rivers.

Western US Prior Appropriation Doctrine

Riparian rights systems in the US were not introduced to the same extent in all parts of the US. The riparian doctrine was well developed in the humid Eastern region. However, in Western states like California where the debates between the prior appropriation and the riparian systems were the liveliest, a mix system of law was eventually adopted. In fact, the relationship between use- and land- based rights systems is not necessarily conflicting. However, half of the Western states completely rejected the riparian doctrine and developed the so-called 'Colorado doctrine', a strict prior appropriation system[10]. The riparian doctrine was criticized for compromising the beneficial use of waters, for limiting water use to riparian landowners only, for being uncertain and inefficient because of their vague and relative character, and prejudicial to the public interest.

These critics reached the Eastern regions later in the 20th century when water uses had significantly increased. Many Eastern states modified the riparian rights

---

[8] 'Pothier lays down the rule very strictly, that the owner of the upper stream must not raise the water by dams, so as to make it fall with more abundance and rapidity than it would naturally do, and injure the proprietor below. But this rule must not be construed literally, for that would be to deny all valuable use of the water to the riparian proprietors. It must be subjected to the qualifications which have been mentioned, otherwise rivers and streams of water would become utterly useless, either for manufacturing or agricultural purposes. The just and equitable principle is given in the Roman Law: Sic enim debere quem meliorem agrum suum facere, ne vicini deteriorem faciat' Kent Commentaries p 354, 355 (1828) quoted in Wiel 1919, 138.

[9] Anyone may improve his own land provided he does not do it in a way that makes his neighbour's land worse.

[10] Arizona, Colorado, Idaho, Montana, Nevada, New Mexico, Utah, Wyoming.

system from the 1950s on with the introduction of permit systems administered by a state water authority. In fact, fourteen Eastern states modified their common law water rights doctrine by introducing prior appropriation elements. Administrative permit systems in the East incorporated prior appropriation elements such as 'beneficial use' requirements and the 'use it or lose it' rule. These prior appropriation elements are however very different from the prior appropriation rights system applied in the West. Contrary to the West, in the East, water rights are attributed for a limited duration, permits are usually linked to the land and cannot be transferred. In the East, water legislations often provide for a particular priority order in case of shortage which does not follow a strict temporal priority order.

The development of the prior appropriation system in the American West in the 19th century has been well documented and described as a 'revolutionary' and dramatic change in water right doctrine (Scott and Coustalin 1995). Scott and Coustalin identified three groups of settlers who particularly pushed for the development of this new doctrine: California gold miners, mountain-state open-country settlers and ranchers, and land developers and governments. The Homestead Act provided land rights for those who cultivated it. Because the river banks were rarely suitable for cultivation and because of the arid conditions of the West, it was necessary to transfer water and irrigate land. Contrary to the East, in the West, irrigation was the most important use of water. To suit the particular historical and geographical conditions of the West, and also in the absence of government and court to apply the common law regarding water, the prior appropriation system – sometimes described as a self-help system- developed in the West. This system is often resumed in two rules: 'first come, first serve' and the 'use it or lose it' rule.

The first appropriative water rights status was introduced in California in 1872. The 1872 California civil code mentions:

> 'A person desiring to appropriate water must post a notice in writing, in a conspicuous place at the point of diversion stating. . .[an amount claimed, purpose, proposed means of diversion, etc.] A copy of the notice must be [. . .] recorded in the office of the recorder of the county in which it is posted' (Lasky 1929).

The three main characteristics of the Western appropriation system is seniority, beneficial use and transferability. In this system, water rights are absolute, not relative. In case of shortage, senior rights have priority over more recent rights. Rights exist in perpetuity as long as water is put to beneficial use. Water rights are not linked to a particular land, meaning they can be transferred.

California continued however to recognize riparian rights along with prior appropriation rights. This mix system or dual system – also called the 'California doctrine'- was applied in Washington, Oregon, California and Dakota, Nebraska, Kansas, Oklahoma and Texas[11].

Besides, California water law was influenced by the 'pueblo rights' doctrine. Indeed, California was ceded to the US by Mexico in 1848, and the Treaty of Guadelupe Hidalgo confirmed the property rights then existing under Mexican law. Under Mexican law, water rights were held by municipalities (pueblos) on behalf of their inhabitants.

Thus, two formal doctrines have dominated the Western world: the prior appropriation and the riparian doctrine and some combination of both. We also tried to show that there exist different versions of these two doctrines. However, many other systems of water rights are used either as a complement to these formal doctrines or as alternative systems. Customs rules, religious law, but also geographical and climatic conditions are the foundations of a large variety of local – often self-organized- water management systems. Water right systems are as diverse as the world from the bisse system in the Swiss mountains to the paddy irrigation schemes in Indonesia or the Philippines and from the Nepalese self-organized farmer groups to Saudi Arabian Islamic water management systems.

THE RECOGNITION OF WATER AS AN ECONOMIC GOOD IN CURRENT DOMESTIC WATER REFORMS

Many states have recently adopted or amended their national water legislation in an effort to respond to today's water management challenges. The four Dublin Principles represent general guidelines for water management reforms. Most current reforms generally attempt to implement these principles. The most controversial principle however, as already mentioned, is that of water as an economic good. For these reasons, our brief review of current trends in domestic water legislation will essentially focus on the recognition of water as an economic good and show its consequences for international watercourses management.

Formalization of Water Entitlements

Today one observes a general evolution toward the creation of formal water rights or permit systems. The formalization of water rights was one of the central as-

[11] Trelease mentions : 'In the meantime, the courts and legislatures of these states and territories adopted common law principles of riparian rights as applicable to privately owned lands often without any thought of conflict with the doctrine of prior appropriation and without any conscious choice of one type of law on the other' (Trelease 1954, 29).

pects of the following water reforms: 2004 in Indonesia, 2002 in Kenya, 1998 in Madagascar, 1992 in Mexico, 1998 in South Africa, 1998 (and its amendment in 2002) in China, 1998 in Vietnam, 2004 in Australia, 1999 amendment to 1985 law in Spain, Brazil (1997-2000).

As shown in the former section, water rights have always been evolving. Today's main response to the growing pressures on water resources is the creation of formal water entitlements and the clarification of these entitlements.

Usually, licences and permits for the use of water are first introduced as a way to register the various uses and estimate the overall water uses in the society. In the first phase, they are essentially used as a tool by water planning authorities to make an inventory of the different water uses. But, over time, these forms of entitlements are more carefully delivered, monitored and enforced. The rules surrounding the delivery of these rights become more complex.

There exist very different systems of water entitlement. Licences and permits usually attach a number of conditions to the use of water or discharge of wastewater. Generally speaking, some uses, such as domestic uses, do not require a licence or a permit. Licences are often purpose-specific, linked to a particular location and limited in time. Licences do not concern consumptive uses only but also 'non-consumptive' uses, such as for the production of hydropower. Orders of priority are also often defined in the case of a water shortage. Some national legislation, such as that of Germany (1986), provide for the possibility to impose new conditions on a permit already granted, such as new environmental requirements (article 5). Water permits can be revoked for various reasons by water administrations.

In all cases, the creation and refinement of water rights imply the development of complex administrative systems to monitor, regulate, and enforce these entitlements. An important part of today's legislation has put into place decentralized management systems corresponding to river basins and sub-basin delimitations rather than other administrative delimitations (Brazil 1997, Morocco 1995, China 2002, South Africa 1998, etc.).

Water Licensing in the Inkomati Water Area, South Africa

In South Africa, the 1998 National Water Act created Catchment Management Agencies (CMAs) (section 77 of the 1998 NWA). These agencies are responsible for the preparation of water allocation plans and the issuance/monitoring/enforcement of water use licences. The first of the 19 Catchment Management Agencies to be created was located in the Inkomati Water Management Area (WMA) in the early 2000s. The Inkomati Water Management Area is situated in the north-eastern part of South Africa at the borders of Mozambique and Swaziland. Thus, an important challenge facing the catchment authorities was to design rules for licensing issuance in an area where the most important rivers are transboundary rivers. The main rivers in the Inkomati WMA are:

- the Komati River (also called Inkomati or Incomati), which rises in South Africa, flows through Swaziland and then re-enters South Africa before flowing into Mozambique,
- the Crocodile River (East),
- the Sand/Sabie River,
- the Uanetse River that flows through the Kruger National Park into Mozambique;

This example clearly illustrates the challenges raised by the formalization of water entitlements on one side of the border in a transboundary basin. How to define the amount of water licences? Water planning in the Area is complicated by the international character of the rivers. In a paper reviewing the experience of the Inkomati Catchment Management Agency, Phil Woodhouse and Julia Brown (2004) report the comments of the Department of Water Affairs and Forestry (DWAF) chief engineer of the Mpumalanga province who recognizes that the 'DWAF should not have tackled the Inkomati first because it is the most controversial and complicated WMA in South Africa'. They present in great detail the lack of an institutional framework to coordinate the decisions of the catchment management agency with the larger river basin. The Komati Basin Water Authority (KOBWA) was created in 1992 by a bilateral treaty between the Kingdom of Swaziland and the Republic of South Africa essentially to operate two dams (the Driekoppies Dam in South Africa and the Maguga Dam in Swaziland), but does not provide any clear allocation of the river water between the two states.

It seems that the creation of the WMA in the Inkomati and the beginning of the licensing process triggered the renegotiation of the institutional arrangement in the basin. Given the particular geographical and hydrological conditions of the region, the negotiation of an agreement over the Inkomati water uses was linked to the negotiation over the Maputo River. Thus, one can acknowledge the role of the implementation of the South African water reform - including the formalization of water rights - in the acceleration of the negotiations between South Africa, Swaziland and Mozambique over the uses of the two rivers that cross their territories: the Inkomati and the Maputo Rivers. The Tripartite Interim Agreement on the Projection and Sustainable Utilization of the Water Resources of the Inkomati and Maputo Watercourses (Interim InkoMaputo Water Use Agreement) was signed between the three countries on the margins of the World

Summit on Sustainable Development held in 2002 in Johannesburg, South Africa. The Agreement stipulates the allowable water withdrawals in the three basin states. Thus, the Inkomati Water Agency has to take into account the limitations on the consumptive water use from the Inkomati watercourse established by the Agreement[12]. South Africa is entitled to:

Domestic, livestock and industrial water use	183 million m ^{3}/annum
Irrigation water use	381 million m ^{3}/annum
Afforestation Area	90 233 ha
Runoff reduction	99 million m ^{3}/annum

The interim agreement provides that two other more comprehensive agreements will be signed after the completion of detailed joint basin studies (expected in 2010–2011). Indeed, the interim agreement requires further negotiations on the development of a common institutional framework and common operating rules to monitor and enforce the withdrawal limitations. As we will see later, the negotiation of such an institutional framework encompasses many challenges.

South Africa is just one example of the consequences of domestic water reforms and the formalization of water entitlements on international rivers. These cases will become more numerous in the near future as domestic reforms are more carefully implemented and water use increases. In other words, in the same way that domestic water law evolved from riparian rights toward administrative water use rights system, it is expected that international water law will evolve from equitable and reasonable use principle towards a more formalized system of international water use rights.

As we will see later, this evolution needs to be accompanied by a clear definition of the property regime of the watercourses. This is a real challenge for international watercourses management.

Water Pricing

Pricing policies are another aspect of current domestic water reforms that can bear important consequences for international watercourses management. Recognizing the economic value of water resources, many national water laws provide pricing mechanisms for water withdrawals, uses and wastewater disposal. But water pricing policies may have different- and sometimes conflicting- goals

[12] The Tripartite Interim Agreement on the Projection and Sustainable Utilization of the Water Resources of the Inkomati and Maputo Watercourses.

such as economic development, public health, water conservation, water reallocation, or social equity. The South African Water Act presents in the following manner the various objectives of water use charges:

> 'Part 1: Water use charges
> In terms of Part 1 the Minister may from time to time, after public consultation, establish a pricing strategy which may differentiate among geographical areas, categories of water users or individual water users. The achievement of social equity is one of the considerations in setting differentiated charges. Water use charges are to be used to fund the direct and related costs of water resource management, development and use, and may also be used to achieve an equitable and efficient allocation of water. In addition, they may also be used to ensure compliance with prescribed standards and water management practices according to the user pays and polluter pays principles. Water use charges will be used as a means of encouraging reduction in waste, and provision is made for incentives for effective and efficient water use.' (1998 South African National Water Act, chapter 25)

The main criteria generally used to establish water prices are the following: the quantity and place of the water withdrawn, the type of uses, the type of crops cultivated, the type of technology used, the type of sources from which water is abstracted, but also sometimes the 'profit' made by the user.

In France, for example, water prices are defined by each water basin agency following a national rule. Charges are supposed to reflect maintenance cost, administrative cost and environmental cost. The formula used to calculate water charges is the following:

$$\text{Charge} = Q*U*S*E$$

Charge is the annual amount to be paid by the user; Q is the quantity of water consumed or pollution generated; U is the unit charge value; S is a sectoral coefficient; E is an environmental coefficient.

Thus the French system introduced in 1964 differentiates between zones in such a way that takes into account the scarcity value of water and the need for investment in regulation and other necessary works. The sectoral coefficient enables a distinction between various user sectors while the environment coefficient reflects the receiving body's capacity to absorb pollutants. The 2006 water law allows also for water prices variation during the year in touristic *communes* with high seasonal population variation.

The contribution of the agricultural sector to water taxes, especially pollution taxes was one of the most important issues in the debate over the adoption of the 2006 French water law. Article 84 dealing with taxes on pollution (§2: '*redevances pour la pollution de l'eau*') increases the price to be paid by the agricultural sector but in a very moderate manner only. The contribution of the agricultural sector to the agency's pollution tax increased from 1% to 4% with the 2006 law. The consumers are still the major contributor to the pollution taxes with 83% of the *redevances* (industry: 13%). Therefore, the polluter pays principle is not thoroughly applied in France.

In most European countries, taxes on water pollution have been increasing in the last decade to reduce all types of water pollution. In Germany, for example, charges for pollution initially covered only organic and suspended matter. Salinity and toxicity were introduced in 1973 and 1974, nitrogen and phosphorus in 1982, hydrocarbons and other inorganic materials in 1992 (Solanes and Gonzalez-Villarreal 1999). For the same reasons as in France, rural users are largely exempted from this system.

The 2000 EU Framework Directive on water provides clear guidelines for water pricing. Article 9 of the EU Framework Directive provides that:

> 'Member states shall take account of the principle of recovery of the costs of water services, including environmental and resource costs, having regard to the economic analysis conducted according to Annex III[13], and in accordance in particular with the polluter pays principle. Member states shall ensure by 2010
>
> - that water-pricing policies provide adequate incentives for users to use water resources efficiently, and thereby contribute to the environmental objectives of this Directive,
>
> - an adequate contribution of the different water uses, disaggregated into at least industry, households and agriculture, to the recovery of the costs of water services, based on the economic analysis conducted according to Annex III and taking account of the polluter pays principle.
>
> Member states may in so doing have regard to the social, environmental and economic effects of the recovery as well as the geographic and climatic conditions of the region or regions affected' (Article 9, EU Directive 2000/60/EC on Water Policy, 2000).

---

[13] Annex III- Economic Analysis : 'The economic analysis shall contain enough information in sufficient detail (taking account of the costs associated with collection of the relevant data) in order to:

(a) make the relevant calculations necessary for taking into account under Article 9 the principle of recovery of the costs of water services, taking account of long term forecasts of supply and demand for water in the river basin district and, where necessary:

. estimates of the volume, prices and costs associated with water services, and

. estimates of relevant investment including forecasts of such investments;

(b) make judgements about the most cost-effective combination of measures in respect of water uses to be included in the programme of measures under Article 11 based on estimates of the potential costs of such measures'.

The Spanish water law provides for the payment of fees for the use or occupation of public water. The base value is estimated on the basis of the economic return generated by the asset. Fees are collected and managed by water authorities at the basin level (Spanish law 1985).

In 2005, Brazil's National Water Resource Council (CNRH), an organ linked to the Ministry of the Environment, unanimously approved a resolution determining the collection of a fee for the use of bulk water resources. Rates are to be set by regional hydrographical basin committees, according to local needs.

Integrated Water Pricing and International Watercourses Management

Peter Rogers, De Silva and Bhatia (2002) underline the importance of 'integrated' water resource pricing. Pricing policies need to be part of a whole system of regulations including all types of water resources in order to be efficient. Indeed, for example, if only the withdrawals of water from rivers are charged, users will progressively turn to another type of water resources that is not charged or not regulated, which is often the case of groundwater. This is true for water licensing as well as for water pricing. The formalization of water rights and water pricing policies are demanding, and require that all types of water be regulated. If not, it will lead to unsustainable outcomes. Peter Rogers *et al.* note:

> 'Industries' decision to buy from public water suppliers, directly abstract, or treat and reuse depends on many factors. Among these are water quality required by that industry, whether the public supplier charges full-cost from industrial users, and whether direct abstraction licences whose charges reflect the full economic cost of water are available. If one or more water source components is weakly regulated, industries will exploit that weakness to minimize their production costs. Due to the common-property nature of groundwater and wastewater disposal services firms are likely to ignore social costs unless they are reflected in effluent taxes or extraction taxes' (Peter Rogers, De Silva and Bhatia 2002, 12).

Thus, because water pricing and water licensing require a holistic approach including all types of water resources, these policies will necessarily have consequences for international watercourses management. Basically the implementation of these reforms will increasingly trigger the clarification of water entitlements on international watercourses and aquifers.

Water prices are essentially determined by more or less advanced administrative systems. Mainstream economists denounce however administrative pricing mechanisms for not reflecting the 'true' cost of water and for not taking into account the efficiency of the different water uses. According to them, the true cost of water should include the opportunity cost of water that is the value of water in its next best use, in such a way as to allow a reallocation of the resource to its highest valued-use. In other words, water markets are preferred to administrative water pricing and are increasingly presented as models for water reform worldwide by some influential mainstream economists.

Trend toward Tradable Water Rights and Water Markets

Trade in water rights and water markets are not a new idea. Some American economists presented the merits of water markets as early as the 1950s and 1960s (Milliman 1959, Hirschleifer, De Haven, and Milliman 1960, Gaffney 1962). Besides, some territories like Chile or the Canary Islands have been experiencing more or less formal systems of water markets for decades. But today, formal water markets are increasingly becoming a reality in some states. Considering the general neoliberal trend of the world economy, the movement towards the creation of water markets will probably strengthen in the years to come.

One observes today that water markets are receiving increasing attention as a useful tool to improve water management and promote efficiency in the use of the resource. Beyond the influential publications of the World Bank, many other authors recognize with more or less resignation water markets as the next step in water reforms. For example, Hodgson, in a 2004 study for the FAO mentions:

> 'Nevertheless, at the present time, it seems quite likely that trade in water rights will increasingly be permitted by water legislation so as to provide existing right holders with an incentive to conserve water and to make more water available for cities in a way that does not involve the construction of expensive and often environmentally controversial dams, canals and pipelines. Ultimately, from a political perspective the introduction of tradable water rights will probably offer a relatively less painful and probably more efficient way of re-allocating water from less productive to more productive uses. So what will the implications be?' (Hodgson 2004).

In a recent study on the 2004 Indonesian water reform, the authors recognize:

> 'Given the interest of numerous international institutions in seeing water use rights made transferable, it is likely that the issue of water use transfers will be revisited in Indonesia's future' (Suharto Sarwan, Tjock Walujo Subijanto and Charles Rodgers 2005, 258).

In 2005, the World Bank released an important report on *India's Water Economy* in which the authors clearly set the stage for the establishment of formal water entitlements separated from land and the introduction of water market systems. This report is a good illustration of World Bank's water policy in the world today. They refer to the existence of numerous informal water markets as an implicit reason for the introduction of well-functioning water markets.

> 'In India, as in all parts of the world where water is scarce, informal water markets have arisen, in which those who have (implicit) rights sell water to those who need it' (Briscoe 2005, 60).

Referring to the establishment of formal water rights, they state:

> 'Once established, such entitlements give rise to a series of fundamental and healthy changes. First, those requiring additional water (such as high-value agriculture and people living in growing cities) will frequently be able to meet their needs by acquiring the entitlements of those who are using water for low-value purposes'(Briscoe 2005, 60).

Further, they explain:

> 'the last ten years has seen enormous progress globally in the use of formal water entitlements- with well-functioning systems now working in Australia, Chile, Mexico, Argentina and South Africa. It is noteworthy that all such established systems are working, often after initial adjustments, and are performing well. In none of the countries that have adopted such systems is there any thought to returning to the previous government-managed allocations procedures' (Briscoe 2005, 62).

If there is indeed a general trend toward water markets, it is important to mention also the existence of very different forms of water markets. In the next chapter we will develop the various types of water markets that exist and discuss in the third part of this book the consequences of the introduction of water market systems on international watercourses management. As we will see later, the main source of differences among water markets is the way water property regimes are defined.

In any case, the formalization of water rights, water pricing and water markets are all illustrations of the recognition of water as an economic good in domestic water policies. It should be strongly highlighted here that these domestic policies are usually accompanied by the recognition of water resources as public ownership. The incorporation of all types of water resources in the public domain might take different forms in different countries but it is a fundamental element in current water reforms.

Water Resources' Incorporation in the Public Domain

A necessary correlative measure to the development of formal water entitlements, water pricing and water market has been the growing incorporation of water resources into the public domain of states. In many states the distinction inherited from Roman law between private and public waters has been erased.

State ownership and control of water resources usually include surface water but increasingly groundwater, desalinated water, rainwater and wastewater as well. The

Spanish water law is a good illustration of the efforts to incorporate of all types of water into the public domain. Recalling the principle of the hydrological cycle, the 1985 water act provides that all continental waters- including rainwater- belong to the 'public hydraulic domain'. In 1999 the public hydraulic domain was extended to 'water resulting from desalination of seawater once it has left the production plant' (Act 46/1999 of 13 December). In Australia and in South Africa, for example, the collection and storage of rainwater that flows on an individual property is regulated. The volume of water tanks and water reservoirs on personal lands are limited.

The 1959 Israel water law provides that 'water resources in the state are public property' (article 1). And article 2 defines water resources very largely:

> Art.2: 'For the purposes of this Law, 'water resources' means springs, streams, rivers, lakes and other currents and accumulations of water, whether above ground or underground, whether natural, regulated or made, and whether water rises, flows or stands therein at all times or intermittently, and includes drainage water and sewage water'.

The control of the government on all types of water resources in Israel are further illustrated by article 4 which stipulates that:

> Art. 4 : 'A person's right in any land does not confer on him a right in a water resource situated therein or crossing it or abutting thereon'(1959 Israeli water law).

The incorporation of water resources in the public domain took different forms such as the declaration of water resources as public property, state ownership, trusteeship for the people of the nation or common heritage of the nation.

For example, the Nepalese 1992 water legislation mentions: 'The ownership of the water resources available in the Kingdom of Nepal shall be vested in the Kingdom of Nepal' (article 3).

The 1992 French water legislation declares water as a 'common heritage of the nation' and recognizes the public interest in the management of the resource[14].

Another example is the 1998 South African National Water Act which declares the public trusteeship of the nation's water resources:

> '1- As the public trustee of the nation's water resources the National Government, acting through the Minister, must ensure that water is protected, used, developed,

[14] Article 1 - *'L'eau fait partie du patrimoine commun de la nation. Sa protection, sa mise en valeur et le développement de la ressource utilisable, dans le respect des équilibres naturels, sont d'intérêt général' (loi française de 1992 sur l'eau).* Water belongs to the common heritage of the nation. Its protection, its utilization and the development of its useable part, in the respect of natural balances, are of general interest.

conserved, managed and controlled in a sustainable and equitable manner, for the benefit of all persons and in accordance with its constitutional mandate' (Article 3, South African National Water Act, 1998).

The level of publicness ownership of water resources depends then on the type of institutions and decision-making procedures established (for example, water tribunals, water police, water user associations and participatory water planning).

If the common property regime of water resources is usually well recognized in domestic water legislation, this is not the case for international water resources.

When domestic water legislation addresses the topic of international watercourses it is usually to reaffirm national sovereignty over these types of resources. The 2001 Kyrgyz law on interstate use of water bodies declared: 'Water resources of Kyrgyzstan are its national heritage' and further made an implicit recognition of state's ownership of international waters 'within its state borders.'[15]

But generally, the property regimes of international watercourses are neither addressed in national water legislation, nor in international treaties leaving unsettled the status of transboundary water resources. More importantly, the institutional framework over international watercourses and aquifers is often too weak to implement a common property regime. The question of the property regime of international watercourses is a central question in our work and will be developed later on.

The large majority of current domestic water reforms is increasingly recognizing the economic value of water resources in three broad ways. Firstly, in order to face growing pressures on water resources, many governments have introduced or clarified formal water rights. They have secondly established or refined pricing policies. And today water markets and trade in water rights are introduced with more or less success in various countries.

These policies might have consequences on the way international watercourses are managed. In the same way that domestic water law evolved from riparianism toward administrative systems of water entitlements, international water law will probably increasingly move away from the principle of 'equitable and reasonable use' toward more formalized systems of water rights. The question of the property

[15] Article 3 : Fundamental Principles of Cooperation in the Field of Interstate Water Relations Implementing the state policy on the use of water resources of the rivers originating in Kyrgyz Republic and flowing outside its borders, and participating in any interstate negotiations on water issues, Kyrgyz Republic acts on the basis of the following principles and terms: recognized right of the state for the ownership of water bodies, water resources and water structures within its state borders; recognized nature of water as a natural resource and economic good which has its economic value with all competing water uses; [. . . .] payment for water use in interstate water relations; rules and amounts of payments for interstate use of water resources of Kyrgyz Republic are regulated by the agreed rates considering the world prices and water demands;

regime of international watercourses will then become crucial. Indeed, in domestic politics, the recognition of water as an economic good (in a broad sense) has been accompanied by the incorporation of water resources in the public domain. Institutional frameworks and decision-making procedures have been introduced in order to guarantee the public nature of water resources property regime. The recognition of water as an economic good will lead to unsustainable, inefficient and inequitable outcomes if it is not accompanied by a strong institutional framework enforcing the common property regime of international watercourses. What does it mean to manage international watercourses and aquifers as common property? This question is much more difficult to answer when it concerns transboundary watercourses than national water resources. How to coordinate the different levels of public interest involved in an international watercourse? Who and according to which principles will design and implement these policies?

Chapter 3
Water Market Systems

The following chapter is devoted to the study of water markets. Water markets are indeed increasingly presented as *the* model for water reforms worldwide. Among the various policy options recognizing water as an economic good, water markets are the most influential. As we already mentioned, because water markets are supposed to reveal the opportunity costs of water resources (scarcity value), mainstream economists tend to prefer this type of allocation system over others.

Water markets are often defined as 'a reallocation of water in which the value of the water is considered independent of the value of land and improvements, buyers and sellers participate voluntarily, and terms of the sale, including the price, are decided by the buyer and seller' (Saliba and Bush 1987, 3–4). In the present work, water markets are defined in the following manner: a water market is an institutional design that facilitates the reallocation of water among different uses through voluntary exchanges between buyer and seller of water rights. Water markets include various types of transactions such as temporary or permanent transfers and spot or asset exchanges. This definition includes water banking schemes.

Basically, the main element of distinction among the various water markets presented below is the degree of government involvement. Administrative procedures are more or less constraining and transfers are more or less regulated.

ISBN: 9781843393610. Published by IWA Publishing, London, UK.

In other words, the main difference between the various types of water markets is the way the property regime of water resources is defined and implemented.

We will first study the characteristics of the Chilean water markets as it is often presented as *the* example for water reforms (Briscoe 1996).

THE CHILEAN WATER MARKET

Chile experienced numerous water reforms with its first code introduced in 1855 (1855, 1951, 1967, 1981, 2006). Water resources are currently regulated by the 1981 Water Code. Under this code, a holder of a water right (*derechos de aprovechamiento*) is legally entitled to use the right as if it were a privately appropriated good. Applicants for water rights do not need to specify the intended water uses. Water rights are granted upon request to the *Direccion General de Aguas* (DGA) free of charge. Article 12 of the 1981 Water Code distinguishes between different types of uses: consumptive or non-consumptive, permanent or contingent, continuous, discontinuous or alternating uses. If there is not enough water to satisfy simultaneous applications for new rights, DGA holds a public auction and sells the new rights to the highest bidder. Application for water rights must be published in local newspapers and the public has thirty days to oppose the new entitlement. All water rights must be registered by the Real Estate Titles Office. These rights have the same constitutional guarantees as are provided for property in article 19–24 of the Constitution. Water rights are perpetual. There is no requirement of beneficial use. There is no regulation to protect either third party interests or the environment. The Water Code establishes no legal priorities among different types of water uses. And, until 2006, water rights owners did not have any legal obligations to use their entitlements, which was a door open to important speculation in water rights. In 2006, a reform was introduced requiring water rights applicants to justify the amount of water they requested and imposing a tax on those who do not use their water allocation (Business News Americas, 13 May 2005).

Water rights are fully divisible and transferable, like other forms of real estate. Usually water transfers do not need to be approved by the Direccion General de Aguas except for transfers using a natural channel and transfers outside a Water User Association (WUA). The price of water rights is not subject to control by a regulatory agency in Chile.

Basically, water rights are governed by private law rather than public law. In fact, water management in Chile results from private bargains among water rights owners and civil court decision.

The Chilean 1981 Water Code, the most liberal water legislation in the world, has often been presented as a success and model for international water reform

(World Bank's and other mainstream economists). Carl Bauer who is a specialist of the Chilean water market presents more prudent conclusions. For Bauer, the Chilean water market is in fact not very active, except in the Limari River Basin. Among other things, the inactivity of the market is explained by the inadequacy of existing infrastructure, canals, and reservoirs, which do not allow many transfers of water. He explains also that there is a legal confusion between customary rights, recognized rights by past legislation and formally registered rights under 1981 Water Code. He mentions that the great majority of water rights in Chile are in fact not formally registered.

Finally, he points out Chile's difficulty to reform its own free water market. He writes:

> 'countries and governments should not make the mistake of thinking that they can implement reforms in two steps, by first adopting a free-market approach to water economics as a straightforward initial step, and then turning their attention to the remaining problems of integrated water resources management and water governance. At that later point their hands will already be tied by a definition of property rights that has major political and institutional implications' (Bauer 2004, 135).

Thus, although the Chilean Constitution defined water resources as a national good for public use, the actual management of water resources is in practice almost entirely done by private owners of water rights. Water use rights are so large and unrestricted that they correspond almost to a private property.

WATER MARKET IN THE CANARY ISLANDS

The Canary Islands are volcanic islands, generally arid with little rainfall. Underground water is the main source of water supply. No regulations clearly define the property regime of groundwater. In fact, groundwater belongs to the one who paid for the infrastructure extracting the water. The construction of horizontal galleries and wells generally requires large investments, leading farmers to create *communidades de agua* (water communities) to share the investments costs. Shares in *communidades de agua* are freely marketable. The Canary Island markets, essentially those of Gran Canaria and Tenerife, are composed of three types of market: water rights market, spot water market and shares of pipes market (infrastructure).

In Tenerife, water and shares are sold by intermediaries in a square in Santa Cruz. The market is recorded to be very active. In fact, the water grid facilitates transfers of water and water rights. Water markets have allowed the transfer from the agricultural to the tourism sector.

Although these markets have been presented as 'successful' water markets by World Bank economists (Simpson and Ringskog 1997), other studies show another picture of these market systems (Aguilera Klink and Sanchez Padron 2002). These authors show how the digging has led to the disappearance of public sources and springs of surface water. They mention that there is no control of groundwater withdrawals and that the water resources of the island (Tenerife) are significantly shrinking. Besides, they denounce the monopoly of the main owners of transportation networks who hold a strong position on the market. They charge important fees for carrying water between parties although pipes are generally leaking and not equipped with meters. This situation has led to the construction of parallel networks of individual pipes, where it was possible, creating complete anarchy in the pipe system.

In 1990, a water law was adopted to promote the sustainable management of the aquifer and created the *Consejo Insular de Aguas* in charge of establishing water management plans. However, it was decided that groundwater concessions will remain under full private ownership for the next 75 years (Ley de Aguas, 1990 article 79). Thus, the public sector has started to invest massively in alternative sources of supply such as wastewater recycling and desalination. Concluding its study on the water market in Tenerife, Aguilera mentions that 'it is not the market mechanism per se that is failing but the institutional framework in which the mechanism is inserted'[1] (Aguilera Klink 2004, 9).

WATER MARKETS AND WATER BANKING IN THE AMERICAN WEST

General Overview

Although there is a large literature on water markets in Western USA, water markets per se are generally not very active in the West. In this literature, water markets are understood in a very broad sense including infrequent but large-scale 'water transfers'. In fact, one notices huge differences in the pace in which water markets are implemented in the West. In the 1990s many states introduced water banking mechanisms to facilitate water reallocation essentially in periods of shortage. Active markets such as those of the Lower Rio Grande Valley in Texas and the Northern Colorado Water Conservancy District are, for the time being, the exception rather than the rule.

[1] 'En calquier caso, no es el mecanismo de mercado el que falla, sino el marco institucional en el que ese mecanismo está inserto' (Aguilera Klink 2004, 9)

As mentioned above, the Western system of prior appropriation separates water from land, enabling the transfer of water to other users, uses and places. However, in some states with a dual system of water rights, like California, only prior appropriation rights are transferable. Riparian rights and correlative groundwater rights are not separable from land. In any case, water transfers are neither new nor frequent in the region. The lack of enforcement, the evolving definition of water rights and the general lack of infrastructure are some factors of explanation of the small number of water transactions.

However, large scale transfers such as the one linking the Owens valley to the city of Los Angeles has drawn national and international attention. From 1905 to 1935, the Board of Water and Power Commissioners of Los Angeles secretly purchased land and water rights from some 869 farms and ranch owners and 825 town lot owners in California's Owens Valley. By 1935, the Agency had acquired 95% of the agricultural acreage and 88% of the town properties in the valley (Libecap 2005). This transfer, operated through violence and bribery has been largely unpopular.

Regulations over water transfers have been progressively introduced, fairly significantly modifying the conditions for transferring water rights. Approval procedures have become more constraining with the introduction of the public welfare or public interest requirements in some Western states' statutes from the late 1890s on. This development is a direct consequence of the declaration of public ownership over water resources in many constitutions and statutes (with the exception of groundwater). Even in states where the public trust doctrine was not explicitly recognized in the statutes, the fact that the statutes or the constitution of that state provided that the water of the state belongs to the public has been considered sufficient to enable the application of the public trust doctrine by courts (*United Plainsmen Assoc. v. North Dakota State Water Conservation Comm'n*, 247 N.W.2d 457, 462–63 (N.D. 1976)).

However, even in state's statutes explicitly recognizing the public interest requirement, there is usually no clear definition of the concept. And the lack of clear definition and criteria to determine the public interest in a transfer approval procedure has been denounced by water market advocates for introducing uncertainty in transfer mechanisms and increasing transaction costs. Many water transfers projects ended in court before judges who developed an evolving doctrine of public trust.

In 1983, the Supreme Court of California delivered a very important decision in which the public trust doctrine was fully integrated into California water law. Explaining that 'the core of the public trust doctrine is the state's authority as sovereign to exercise a continuous supervision and control over' the waters of the state, the Court ruled 'the public trust doctrine and the appropriative water rights system are parts of an integrated system of water law'.[2] Western states' statutes are increasingly including public trust criteria to help the state engineer or other Water Resource Department officers to evaluate new appropriations and transfers applications. The introduction of these reforms raises however sometimes the question of pre-existing water rights. California Supreme Court recognized state power to modify existing water rights as a result of its sovereignty over water resources and its responsibility to enforce evolving notion of public trust[3], in some other Western states, however, the nature of water rights may impede such 'retroactivity'. Depending on the types of water rights (such as groundwater rights and surface water rights) and the state regulating them, water rights will be more or less easily modifiable.

Thus, the formal recognition of water as a public resource in the state constitution or statutes does not necessarily lead to effective state power and control over these resources. The nature of water rights may sometimes clash with state power over the resource.

Besides, in an important decision in 1982, the Supreme Court of the United States referred to state ownership as a 'legal fiction', enlarging thus the scale of 'public interest' to the federal level (*Sporhase v. Nebraska Ex Rel. Douglas*, 458 U.S. 941, 1982) (see the second part of this book).

Before analysing more carefully some institutional aspects of water markets in different Western US states, we found it necessary to raise these general issues surrounding the ambiguous nature of water rights and the true scope of state sovereignty over these rights. Whereas water rights in Chile or the Canary Islands are clearly private property rights, their nature in Western USA is much more complex and depends on the many different regional institutions.

Colorado Big Thompson Project (Colorado)

The most active water market of the West is located in Colorado due to the Colorado Big Thompson Project (C-BT). This market involves both short term water rentals and permanent sales of water rights. Water rights are essentially transferred from agricultural uses to urban uses. The rules applied in the Big Thompson Project are different from those of the rest of the state. The Big Thompson Project is a large water transfer from the western part to the eastern part of the Rocky Mountains. It was supported by the Bureau of Reclamation

[2] National Audubon Society *et al.* v. Alpine County, 33 cal.3d 419.

[3] 1983 Audubon Society *et al.* v. Alpine County, 33 cal.3d 419 'The objective of the public trust has evolved in tandem with the changing public perception of the values and uses of waterways' (p 12).

and was first operational in 1947. Contrasted with water regulation in other parts of the state, in the Big Thompson Project, there is no seniority of water rights. Instead, the C-BT Project provides for a priority order favouring urban, domestic and industrial uses over agricultural uses in case of shortage. Another important rule provides that the Northern Colorado Water Conservancy District (NCWCD) is the only owner of C-BT return flows. Ownership of the return flow was granted to the NCWCD rather than to downstream users. In fact, the NCWCD is the only true owner of C-BT water rights. Finally it is important to underline that the authority of the District over C-BT water was possible because C-BT water was new in the region. NCWCD's authority never clashed with pre-existing claims.

Lower Rio Grande Water Market (Texas)

Behind Colorado, it is the populous states of California and Texas which have the largest number of transactions per year, closely followed by Arizona. The most active water market in Texas takes place in the Lower Rio Grande Valley, a region extending from the Amistad Reservoir to the Gulf of Mexico (see map below). Water rights in the region were adjudicated by a court case in 1971. These water rights entitle their holders to a portion of the combined storage of Amistad and Falcon Reservoirs, the volumes of which are apportioned by treaty between the US and Mexico. Water rights accounts and transfers are administered by the Rio Grande Watermaster (Griffin and Characklis 2002).

The Watermaster is in charge of determining each year the amount of water available for all uses. This amount may vary quite significantly depending on the flow of the Rio Grande, the climatic conditions and the water levels of the Falcon and Amistad Reservoirs. The amount of water allocated to municipalities is determined in priority at the beginning of each year. However, the amount of water allocated to irrigation is determined at the beginning of each month. In fact, the Watermaster administers the accounts of all users which are regularly revised to take into account of the water used and the amount of new reservoir inflows allocated (there are over 813 active water rights along the Rio Grande). Information concerning the water accounts is public, the Watermaster office behave like a broker, bringing buyers and sellers together. Prices are not determined by the Watermaster but by negotiations between the buyer and the seller. The Rio Grande Watermaster also has an important role in monitoring and enforcing water rights. In fact, the enforcement is considered successful in the valley thanks to continuous water patrolling along the river and canals.

Irrigation water rights not used in a particular year may be carried over as storage to the next year. However, the Watermaster defines a storage limit for irrigation rights of 1.41 times the annual water right. Beyond that limit, the excess must be used, transferred or lost. Water rights are cancelled after a 10-year period of non-use (Yoskowitz 1999).

As already mentioned water rights' priority order is not historical but depends on the type of use. Municipal, industrial and domestic uses have priority over agricultural uses. It is recorded that between 1997 and 2002, agricultural users were allocated, on average, only 38% of their paper water rights (Kelly 2004, 35).

The administrative procedures to transfer a water right are simplified. There is no requirement of public notice or third-party impacts consideration. In the region, most transfers occurred from agricultural to non-agricultural uses (90%).

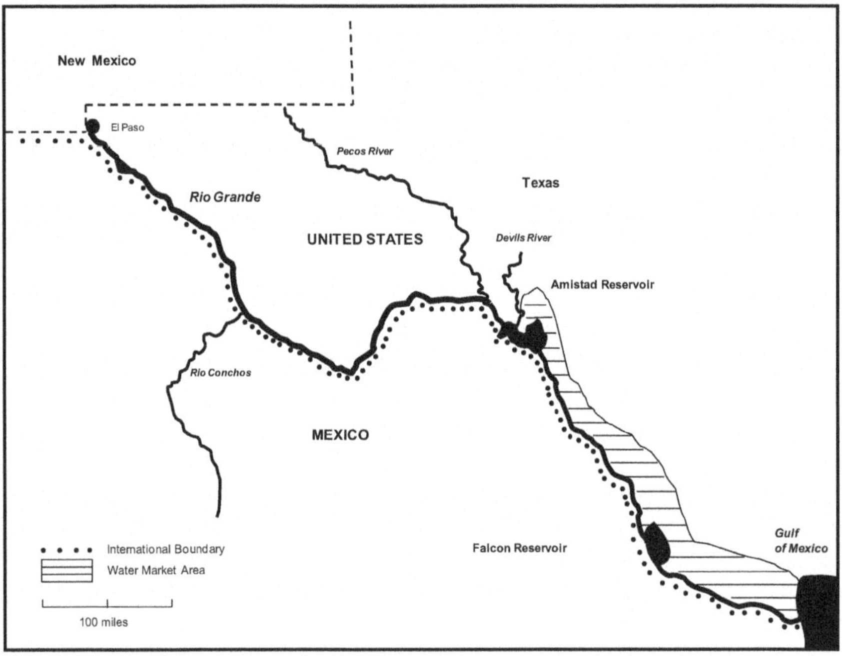

Figure 2 Lower Rio Grande Market Area (after the Texas Commission on Environmental Quality and the International Boundary and Water Commission).

Water Banking

Whereas water markets per se are not very frequent in the West, various types of 'water banking' have developed in the West since the 1990s. A recent study prepared by the State of Washington Department of Ecology has shown that twelve of the eighteen Western states had developed some sort of water banking system (Clifford, Laundry and Larsen-Hayden 2004). The study of the Washington Department of Ecology is based on MacDonnell's large definition of a water bank. He defined a water bank as

> 'an institutionalized process specifically designed to facilitate the transfer of developed water to new uses. The primary objective of a water bank is to bring together those holding legally valid water entitlements interested in making the water available to another with those needing to obtain additional supplies of water for their uses' (MacDonnell 1994).

Thus, water banking is understood in a very broad sense. This broad terminology brushes the distinction between pure water market and water banking. While analysing water banking in Western US, it seems important to keep in mind a broad distinction between water market and water banking. In a water market, buyers and sellers determine prices without an intermediate, whereas in a water banking system, prices are predetermined by the bank who acts as an intermediate between buyers and sellers.

There exist very different forms of water banking mechanisms. Water banking regulations vary with the scale of the water bank (state-wide, regional, inter-state), the types of entitlements to be bought and sold, the quantity of water to be traded and the temporary or permanent character of the bank. Different formats of water banking have been created: so-called 'institutional banking' (paper title), surface storage banking (necessitates physical storage capacity) and groundwater bank.

Generally speaking, a water bank simplifies the procedure for the temporal transfer of water by reducing third-party and environmental impact reviews.

Idaho Rental Pools and State Water Supply Bank

Water banking systems were pioneered in the 1930s in Idaho on the Upper Snake River. MacDonnell explains that this informal water bank emerged in the region as a result of the particular conditions of the Upper Snake River where a large amount of storage capacity was available in dams constructed by the Bureau of Reclamation. As some water districts and water users had storage rights exceeding their direct delivery demands in many years, a banking mechanism was introduced. This banking mechanism served as the model of the first formal water banking in the West, the state-wide water bank established in Idaho in 1979. Although created in 1979, the State Water Supply Bank became active in only 1995. The legislation enacted in 1979 not only created the state-wide water bank but also formalized five local rental pools. Rental pools have rules to determine order of priority for water allocation.

Whereas prices are fixed in each rental pool, in the State Water Supply Bank prices are suggested but not fixed. The State Bank acts as a clearinghouse where buyers and sellers post their intent to buy and sell. Whereas local rental pools have been very active, the State Bank's activity is still sporadic.

California Drought Water Bank

In contrast with the permanent water banking system of Idaho, California established a temporary water bank to facilitate the transfer of water rights during an unprecedented drought in 1991, 1992, 1994 and again in 2009. Before the creation of the 1991 Emergency Drought Water Bank, California had already experienced water banking but only at a local level. The Bureau of Reclamation had established a water bank in the Central Valley for Central Valley Project members during a severe drought in 1977.

The 1991 Water Bank introduced innovative ways to transfer water between and among sectors. This system of water banking sought to facilitate short-term water transfers in periods of water crisis (not a permanent transfer of water rights). It is also a way to soften the 'use it or lose it' rule by allowing temporary water transfers without impairing the corresponding water right. The temporal transfers of water are considered to be beneficial use, and are not subject to loss or forfeiture. Willing sellers can make water available in three main ways: groundwater substitution (using groundwater instead of surface water), reservoir releases (stored water) and farmland fallowing. As mentioned by Howitt, 'the most innovative operations . . . .involved the acquisitions of water from holders of riparian rights who fallowed their lands' (Howitt *et al.* 1992). Indeed, the Bank regulation provided for the acquisition and temporal transfer of riparian rights.

The Bank also provided for an order of priority in the allocation of water resources. Thus, 'extreme critical needs,' including drinking water, health, sanitation and fire protection and urgent agricultural needs, were given high priority. The Bank required that potential buyers implement a satisfactory conservation program before delivering the water.

In 1992, a Drought Water Bank was re-introduced with some modifications. In 1991, the Bank purchased water based on estimates of demand. The Bank bought water at the fixed price of 125$/ acre foot and sold it at 175$/ acre foot at the State Water Project's delta pumping plant at the entrance of the California aqueduct. However, after the Bank made commitments to purchase the water, the weather changed and more rain fell, thus reducing the demand. The Bank was unable to resell all the water it had purchased.

Thus, in 1992, the Bank ruled that buyers had to commit contractually to purchase water before the Department of Water Resources would obtain purchase contracts. The Bank created six separate pools to avoid all purchased water being stored in a single pool, thus reducing transportation costs. Because major controversies were raised by land fallowing contracts in 1991, the DWR decided in 1992 to eliminate land fallowing as a source of water for the Bank. Purchasers were allowed to store purchased water for up to three years (MacDonnell 1994). The 1992 Water Bank was a small bank. Prices were much lower than in 1991 and fewer farmers decided to sell water (purchase price 50$/acre-foot and selling price 72$/acre-foot). The amount of water purchased by the Bank fell from 820,665 a.f. in 1991 to 193,193 a.f. in 1992.

For the first time in 15 years, the Department of Water Resources decided to establish another Drought Water Bank in 2009. Under this program, the DWR bought water upstream of the Sacramento-San Joaquin Delta ($275) and sold it to buyers downstream. The water was transferred using State Water Project (SWP) and Central Valley Project (CVP). As in 1991, 1992 and 1994, willing sellers could make water available through groundwater substitution and reservoir releases. But in 2009, the DWR also allowed cropland idling and crop shifting as sources of water for the Bank. To avoid the controversies raised by the 1991 farmland fallowing program, the DWR decided to strictly regulate cropland idling[4] and limit it to 20% in each county. Although the groundwater substitution program was highly criticized for having contributed to the depletion of some rural drinking-water wells during the 1994 Water Bank, it was reintroduced in 2009. Willing buyers had to make a deposit of $75 per acre foot in advance to be eligible to purchase water from the Bank. The price of water was much higher than in 1991 and 1992. In addition to the water cost ($275 per acre foot), the buyer had to pay the conveyance cost (depending on the delivery location) and administrative costs ($5 per acre foot). As in 1991 and 1922, willing buyers had also to commit to water conservation measures. This time, these measures had to result in a minimum of 20% overall savings. The 2009 Water Bank was also a small bank in comparison to 1991. Only 76,600 a.f. were transferred through the Bank. It seems that some technical, legal and administrative issues deterred some willing sellers and buyers from participating in the Bank. In 2009, various environmental groups sued the DWR for failing to properly

[4] In order to enter in a cropland idling program with the DWR, the willing seller has to document the 'crop history' of the farm or district for the five previous years in order 'to identify the type of crops typically grown and the degree of typical land fallowing that takes place, and typical crop rotation practices.' (DWR 2009, 9)

assess the environmental impacts of its banking activity (particularly on the giant garter snake habitat). As a consequence, some farmers feared they would be liable for environmental damages. Paradoxically, environmental permitting was more complicated than before because of new procedures, including a period of public review. Some willing buyers also feared they would lose their committed money if the DWR could not deliver the water for hydrologic reasons. Finally, rice farmers of the Sacramento Valley had no incentive to sell their water because the rice prices were at their highest levels in 30 years (due in part to the prolonged drought in the Murray-Darling Basin) (Downing 2009).

In the summer of 2009, the DWR was discussing the reintroduction of another drought water bank for 2010 and the extension of the program to all types of transfers from 2011 on. This basically means that transfer procedures will be simplified and harmonized all over the state. Considering water transfers as a standard management tool and not only a drought response, the DWR expected to establish a long-term transfer program from 2011 on.

Texas Water Bank and Texas Trust

The Texas legislature created a water bank in 1993 and a water trust in 1997. The Trust is part of the Bank and is aimed at the acquisition of water rights for environmental and aquatic preservation purposes. The Texas Bank is different from those presented above, in the sense that it is not based on physical storage capacities. Rather, the Bank's role consists essentially in that of a clearinghouse, utilizing a 'bulletin board' approach to match buyers with sellers. Besides, transactions do not involve temporal transfers alone, but also permanent transfers of water rights. The Bank with its Trust may act as a market-participant. The Texas state-wide Water Bank inaugurated in 1994 has remained largely inactive. Yoskowitz advanced many reasons for the lack of activity of this Bank (Yoskowitz 1999). In contrast with the California Drought Water Bank, the Texas Water Bank is permanent. In contrast with the rental pools in Idaho, the Texas Water Bank is state-wide. All kinds of water rights are eligible for deposit in the TWB, i.e. both surface and groundwater rights. Finally, the prices are not fixed by the Bank but are determined by the seller and the buyer. Thus, the Texas Water Bank is potentially very large. Information about the Bank and infrastructure are lacking. Besides, the structure of the market is highly distorted with four percent of water rights holders holding 90 percent of the water rights in the State of Texas. Yoskowitz also recalls that there is no big difference between transferring water rights outside or inside the Bank. He mentions that the only difference between the Bank and the Registry is that surface water rights are protected in the Bank from cancellation for an initial 10 years (Yoskowitz 1999).

Arizona Water Banking Authority

The Arizona Water Banking Authority (AWBA) was created in 1996. Contrasted with other banking systems presented above, the main goal of the Arizona Water Bank is to store Colorado water and ensure that Arizona uses all of its 2.8 million acre-feet of its entitlement to Colorado River water. The AWBA purchases Central Arizona Project water that is not used by CAP contractors and stores it for future state needs. Arizona has recently signed an interstate banking agreement with Nevada to sell its 'intentionally' unused water.

This banking system introduced a new definition of beneficial use. As long as water rights are stored in the Bank either for future needs or for exchange, they are not forfeited. The Arizona Water Bank consists of huge natural and artificial underground storage facilities. The mechanisms of this interstate Bank will be developed in the third part of this book.

Water banking systems are also developing elsewhere in Western United States. Water banking pilot projects have been established on the Arkansas River (Colorado), the Lower Pecos River Basin (New Mexico, 2002) and in the Yakima Basin (Washington, 2003). Oregon created in 2003 the Deschutes Ground Water Mitigation Bank to restore environmental conditions in the Deschutes Basin. Groundwater applicants are required either to purchase mitigation credits from the Mitigation Bank or implement a mitigation project.

AUSTRALIAN WATER MARKETS

Water is managed differently in each of the six states and two territories of the Australian Commonwealth. Water markets first emerged in the eighties in the three states of the southern Murray-Darling Basin. Our analysis will therefore focus on South Australia, New South Wales and Victoria. In sharp contrast with most US water markets, there is no rule of prior appropriation. Water rights have no seniority attributes and all rights for the same-use purpose are treated equally. The federal government took a leadership role in water policy with the adoption of the National Water Initiative in 2004. The 2004 Water Initiative is an agreement among the Commonwealth, the states and territories launching a nationwide water reform process. One of the main objectives of the reform was to harmonize the system of water rights by unbundling water licences into water access entitlements and water use licences. In 2007, the national parliament adopted the 2007 Water Act, the first since 1912, to further harmonize water entitlements and facilitate inter-state trading. State water legislation underwent significant reforms in the last few years and will certainly continue to evolve until the Basin Plan is adopted in 2011.

New South Wales

New South Wales has the largest irrigation industry in Australia. New South Wales is also the first state of the Australian Commonwealth to have adopted a system of tradable water rights. After several years of drought, the government of New South Wales authorized irrigators to trade temporarily their water entitlements on an annual basis. Permanent water right transfers were introduced in 1989.

The current water rights system in New South Wales is regulated by the 2000 Water Management Act. As with the 1912 Act, the Water Management Act of 2000 vests the Crown with the right to the control, use and flow of all water in rivers, lakes and aquifers.

This Act introduces a new order of water uses priority with human needs in first position and environmental needs second, before urban, high security licences and irrigation uses. Thus the 'environment' is recognized as a fully legitimate user of water in this system. Moreover, the Act facilitates water trading by distinguishing between access water rights and use water rights. The regulation recognizes different types of access licences and use rights. The government plays an important role in the reallocation of water resources. Each year, a Water Planning Commission defines how much water will be available to each class of rights and access licences. Water is assigned to each class based on the priority of the access licences. Generally the share (quantity) and extraction (rate) components of an access licence are divisible and transferable. Trade may be permanent or temporary. However, to hold an access licence is not sufficient to use the resource. The licencee must hold a use approval. Use approvals are linked to land ownership or occupancy because they are issued for specific tracts of land and for particular purposes.

Victoria

Water management in Victoria is essentially regulated by the Water Act adopted in 1989. In 2005, the Victorian Parliament adopted a new Water Act to harmonize Victoria water legislation with that of the other states of the Murray-Darling Basin and facilitate interstate water trade (this will be developed in Chapter 8). The new legislation provided for the unbundling of water entitlements from July 2007 on. Until 2007 Victoria did not distinguish between access and use licences.

Under Victoria legislation, as in the case of New South Wales, the Crown has the right to the use, flow and control all water in a waterway and all groundwater. Victoria's system of water entitlements is complex and includes many different types of entitlements such as bulk entitlement, water share and water use licence. The government attributes 'bulk entitlements' in perpetuity to the 'Authorities'. Authorities are 'government business enterprises' responsible for the operation and maintenance of bulk water supply. There are six Rural Water Authorities, fifteen Regional Urban Water Authorities and four Metropolitan Urban Authorities. The Minister appoints the Board of Directors of each of these Authorities. Authorities are in charge of issuing, modifying and transferring water shares on behalf of the Minister. They are also responsible for the monitoring and enforcement of water use licensing system. Although granted in perpetuity, bulk entitlements are at any time revisable under the 1989 Water Act. The Minister has the power to review allocations of any type, including bulk entitlements, and amend these allocations by order. Thus, the Minister may decide to reallocate water between consumptive users and the environment. Bulk entitlements are divisible and transferable to another Authority or to the owner or occupier of a landholding in an irrigation district, to the holder of water share, or to a person outside Victoria. In each case, the approval procedure is different. Transfers may be temporary or permanent. When the 1989 Water Act first entered into force, authorities made use of auction systems to sell part of their entitlements. Indeed, bulk entitlements may be converted into water shares. Authorities can also buy water share and licences and convert them into a bulk entitlement. The Minister is in charge of approving these conversions.

Thus, resource planning in Victoria basically operates at two levels: the government decides bulk entitlements attributed to each Authority, and each Authority is then responsible for the allocation of water shares, all of them being transferable and convertible. In addition to a water share, an irrigator needs to hold a delivery share. A delivery share provides an entitlement to the service of having water delivered to land in an irrigation district at specified volumes during specified periods. Delivery shares remain attached to land within an irrigation district, but are transferable within it. If the irrigator wishes to sell its water share, it can either continue to hold the delivery share and pay the annual access fee, or reduce/surrender the delivery share and pay the relevant termination fee.

The 1989 Water Act also provides the possibility to appeal the Authorities' decision at the Victorian Civil and Administrative Tribunal. Trading activity is reported to be very active, especially in the Goulburn Murray Water District. A website[5] was created to facilitate water share and water allocation (temporary) transactions in different parts of Victoria.

South Australia

The current water rights system in South Australia is regulated by the 1997 Water Resources Act and the 2004 Natural Resources Management Act (as amended in 2007). Contrary to regulations in New South Wales and Victoria, the South Australian Water Act does not explicitly confer ownership of the water resources to the Crown. The Water Act only regulates water *uses*. In fact, the Act distinguishes between two types of water: prescribed water and unprescribed water. Where water is unprescribed, there is no need to obtain a licence to use the water. Riparian landowner, occupier, or a person with lawful access may take water from a watercourse, a lake, or a well for any purpose without restriction as long as their rights do not affect the right of another person from taking water.

Water resources are managed through a hierarchy of Water Plans: the State Water Plan prepared by the Minister, Catchment Water Management Plans prepared by Catchment Management Boards and Water Allocation Plan prepared for prescribed water by either the Catchment Water Management Board and Water Resources Planning Committees, Land and Water Management Plans prepared by local Councils and also Irrigation Drainage Management Plans (to control salinity).

When the Minister announces his intention to prescribe water resources of a particular zone, there is then a period of public consultation and the establishment of a Water Allocation Plan. The Plan contains the criteria by which decisions about the allocation, transfer and use of water are made. The Water Allocation Plan (WAP) is developed in consultation with the community and is adopted by the Minister for Environment and Conservation.

To conform with the 2007 national water reform, South Australia amended its 2004 legislation to allow water entitlements to be unbundled into water access entitlement and water use licence. Until July 2009, all water licences in Victoria were bundled. Before 2009, water licences authorized the holder to take (or to hold) surface or groundwater, subject to particular conditions. Water licences were composed of water holding allocation and water taking allocation. Holding allocations were created as a means to streamline the procedure for water resources transfers. Holding licences were divisible, separated from land title and transferable both permanently and temporarily.

[5] http://www.watermove.com.au

The current water entitlement system is composed of four types of entitlements: water access entitlement, water allocation, water resource works approval (water use licence) and site-use approval (water use licence). Former water licences or today water access entitlements are issued in perpetuity by the Department of Water, Land and Biodiversity. The Department is in charge of the assessment of water licences/access entitlements and transfers applications. This assessment is done in compliance with the provisions of the corresponding WAP. Before 2009, although water licences were issued in perpetuity, the Department could change the allocations and licence conditions at any time without compensation as a result in an amendment to a WAP. In case of a shortage, the Minister could reduce both the water taking and water holding allocations of the licences. Under the new system, it remains to be seen how this will function.

Each WAP has its own specific trading rules but they usually required that water licences/access entitlements be traded downstream in a catchment and not upstream. Because the public is consulted in the elaboration phase of the WAP, there is no public notice of licence/water access entitlement and transfer application (except if they are explicitly mentioned in the WAP). Where a person is noted on the register of water licences/water access entitlements as having an interest in the licence/water access entitlement, the DWLBC must not grant approval for the transfer of the licence/water access entitlement without the written consent of that person. A licensee has a right of appeal to the Environment, Resource and Development Court.

Finally, South Australia has put into place water banking systems in some prescribed areas such as the Angas Bremer Prescribed Wells Area. With this system, a portion of unused water allocation may be 'banked' for further uses. Only up to 30 percent of water licence may be banked. Storage is limited to three years.

MEXICO'S WATER REFORM

Mexico's water reform of 1992 is part of the general liberalization of the country in the early 1990s. The 1992 water reform was preceded by the 1992 agrarian reform which significantly modified the *ejidos* system. *Ejidos* were parcels of land owned by the state, created in 1917 to provide residence and employment. Holders of *ejidos* had usufruct rights only, and were prohibited from renting or selling their parcels. With the 1992 land reform, *ejidos* became private property, enabling the holders of such a title to rent and sell their parcel (Margaret Wilder 2005).

According to the 1917 Constitution, the state has the authority to regulate the exploitation and use of national waters. The Nation is the original proprietor of land and water property which can be granted to private individuals through concessions. The 1992 *Ley de Aguas Nacionales* provides that except for domestic and stock uses, water users must submit an application to the *Comisión Nacional del Agua* (CNA). All waters are regulated except some unregulated groundwater aquifers.

The 1992 federal water law essentially provides for the decentralization of the management of irrigation districts from the federal government to Water User Associations (WUAs) and the introduction of a formal system of tradable water rights. The law provides also for the creation of river basin councils.

Created in 1989, the *Comisión Nacional del Agua* (CNA) keeps a central role in the planning of water resources and the monitoring of the *Programa Nacional Hidraulico* (PNH). It is responsible also for the administration of water concessions.

Water concessions are granted to a variety of entities such as individual users, WUAs, municipalities and other governmental agencies. Water concessions must be registered at the Public Registry of Water Rights. They are issued for a period of 5 to 50 years. The establishment of a formal system of water rights is supposed to provide an inventory and evaluation of current water uses and thus facilitate the management of water resources. However, it seems that in the process of formalization of water rights, water has been over-allocated. Thus, the concessions renewal procedure monitored by the CNA is considered very important in limiting future uses of water resources (Garduno 2005).

Water concessions are separated from land, divisible and transferable. The CNA has however the power to expropriate, fully or partially or to limit the ownership of a water right if it is in the public interest. Article 27 of the 1917 Constitution limits however the power of the CNA by providing protection to private property. Thus any modification or expropriation of a water right must be compensated (General Act Governing National Assets).

The CNA reviews water concession renewals and transfer applications in taking into account the PNH and evaluating third-parties and environmental effects. The CNA may then either grant, deny or set out some conditions for its authorization. New entitlements and transfer agreements are then published in the Official Diary of the Federation. The applicant may appeal the decision under the Federal Code of Civil Procedure (Productivity Commission 2003).

Water concessions are volumetric. In case of a shortage, there is no priority order and no proportional reduction system. Rather, the CNA implements rationing rules, which strengthens its role in case of shortages. Some authors argued that the rationing system reduces the security level of water entitlements (Hearne and Trava 1997).

In each district, the availability of irrigation water is estimated by the CNA at the beginning of the cropping season. This information is made available to farmers who are then required to submit a cropping plan to the WUA. Irrigators' water rights are granted by the WUAs.

Transfer procedures are easier when only the name of the user changes as in the case of water transfer within irrigation districts. In Mexico, informal water markets for seasonal water use have existed for many years. However, the transfer of a concession outside the district requires the approval of the General Assembly of the WUA, as well as authorization of the CNA. The seller of a water right outside an irrigation district does not receive directly the benefits of the sale, rather, it is the district (Hearne and Trava 1997).

WATER MARKETS IN SPAIN

Water markets are not entirely new in Spain. In addition to the particular experience of the Canary Islands, other systems of exchanges of water rights have been functioning for more than a century. But these 'markets' were usually informal and very local. This is for example the case of the Camp de Tarragona water market, including the Siurana-Riudecanyes irrigation district and the city of Reus. Since the beginning of the irrigation project in 1904 and the construction of the dams on the Siurana-Riudecanyes river system, water titles have always been tradable. This water market has been limited to the municipalities and landholders of the region because the ownership of land within the irrigable zone was a requirement for the ownership of a water title (Marino *et al.* 1999).

In 1999, new legislation modifying the 1985 Water Law laid down the basis for the development of formal water marketing in Spain. However, the articles regulating the trading procedures are brief and vague. It seems that the adoption of this new framework was not very publicized and did not trigger much reaction. Article 67 to 72 of the new law (*Texto refundido de la ley de aguas 1985*) provides for two ways to transfer water rights: through water transfer contract between two water right holders (*contrato de cesión de derechos de uso de agua*) and through water exchange centres (*centros de intercambio*). According to the new law, only those who are already water rights holders can enter into a water transfer contract. To own a water entitlement is thus a necessary condition to enter the water market. Moreover there is a strict priority order requirement: the water buyer must be of similar or greater priority than the water seller (article 67.1). What can be traded is not the formal volume written in the water entitlement but the actual volume of water used by the water seller (to be determined as the average volume used in the last 5 years) (article 69). The contract between the buyer and the seller has then to be approved by the basin agency. If the basin agency does not give an answer within two months, the transfer can be considered as approved (article 68.2). During the approval time, the basin agency has the power to buy the water right in exchange for financial compensation as provided in the contract (*derecho de adquisión*, article 68.3).

The new legislation also provides for the possibility of creating water exchange centres under certain circumstances (article 71). In June 2004, the Ministry of the Environment (*Ministerio de Medio Ambiente*) approved the creation of *centros de intercambio de derechos de agua* in overexploited river basins: Segura, Júcar, Guadiana y Guadalquivir. The basin agencies will thus be allowed to bank water rights and sell them later at a fixed price through these centres of water rights exchange. Water banking systems might be implemented in the near future in each river basin on a permanent basis (De Stefano 2005). A further step in the direction of a water market was taken in December 2005 when the Council of Ministers approved a *real decreto* authorizing interbasin water rights transfers in period of droughts. For the time being, only the Segura and Almanzora River basins are affected by this law (Real Decreto-Ley 15/2005, December 16, 2005). In 2006, another modification introduced by *real decreto* allowed the water exchange centres to buy water rights for the environment in order to tackle the severe problem of over-allocation and over-use. However, for the time being, only a few transfers have been recorded. It seems that water market policy in Spain is muddled and hesitant. Furthermore, Spanish water policy clearly has other priorities such as improving water use measurement and registration, designing independent and scientific basin plan and solving the issue of illegal groundwater pumping.

COMPARISON OF THE MAJOR CHARACTERISTICS OF WATER MARKETS

By presenting some of the major water market systems in the world, we aimed to demonstrate the great variety and complexity of the different types of water market mechanisms. We should acknowledge, however, that for the purpose of our study we deliberately left aside market-based solutions addressing water quality problems. These pollution trading systems are essentially limited to the US and Australia[6]. There are, generally speaking, three types of pollution rights corresponding to three categories of water quality problems: salinity, organic pollution (oxygen depleting substances) and nutrient pollution (nitrogen and

[6] e.g. in Australia: the Hunter River Salinity Trading System, the Hawkesbury-Nepean River nutrient pollution rights trading scheme. e.g. in the US: the organic point source trading program on the Fox River (Wisconsin), the Tar-Pamlico River nutrient trading program (North Carolina), the Cherry Creek nutrient trading program (Colorado).

phosphorus) (Kraemer, Kampa and Interwies 2004). These schemes require complex designs and advanced monitoring devices. One of the difficulties of these schemes is to fix a cap on emissions that is not only flexible enough to take into account the specificity of the river flow and variations in assimilative capacities, but also the changing time and location of pollution.

Coming back to the focus of our work, the following criteria can be used to characterize some of the water markets studied. The principal elements of distinction among the various water markets studied are their size, whether or not they are included in a water plan and the level of 'unbundling' of their water rights.

- **Issuance of water rights**
  In most Australian states, applications must be made to the water administration. A water right is attributed only if it corresponds to a river basin plan. In Chile and in some Western US states, they are automatically attributed if their beneficial use does not harm existing water uses.
- **Definition of water rights**
  Water rights are defined very differently in the various water markets studied. Some, such as irrigation rights, are sometimes linked to a particular area (irrigation district) and to a particular purpose (irrigation). Mexico allows water right holders to trade outside their irrigation district, but only under complex procedures. In Australia, water rights have been 'unbundled' into access entitlements (tradable) and use licences (formal use approval, non-tradable). Water access entitlements are defined as a perpetual share of a variable consumptive pool (determined in the basin plan). In the US and Chile, water rights are usually allocated in perpetuity. They correspond to a fixed volume.
- **Water reduction's risk assignment (water rights modification)**
  Australia is the only state to define rules for assigning risks in water reduction. Thus, the risk of reduction in water allocation which arises from climate change or drought is borne by water right holders, but risks resulting from a change of government policy are borne by the governments themselves. Risks resulting from 'a change in knowledge of a water systems' capacity to sustain a particular extraction level' are borne equally by water right holders and the governments. In other places, there is no such distinction. In Mexico, Spain, Chile and the US, water rights usually cannot be modified without compensation.
- **Procedures in case of a shortage**
  Different mechanisms have been established in case of drought: the authoritative reallocation of water between uses (Mexico), the proportional

reduction of all water entitlements, or the establishment of priority orders. In Australia, all water rights in the same category of use have the same priority order. In the US, the system of prior appropriation traditionally provides for a temporal priority order regardless of the type of use (except for basic needs and other highly important uses).

- **Transfer procedures**
  Transfer procedures in the different water markets studied vary in complexity. In Chile and the Canary Islands, there is almost no restriction on water transfers. In Chile, if no one opposes a publicly-announced transfer proposal, the transfer can be executed. In the US, transfers are usually subject to third-party and environmental impact evaluations and are usually open to public review (except in water banking systems were the procedures have been simplified to facilitate transfers). In Spain, transfer proposals should be considered as approved after two months of administrative silence. In Australia, there is no particular impact study as they are included in thc watcr plan. The water plan specifies which type of transfer can occur, in which quantities and in which direction.
- **Environmental protection**
  The environment can be protected in a different manner. In the US, independent agencies, like the Texas Water Trust, participate in the market and buy water rights for the environment. In Australia, allocation for the environment will be provided in the basin plan and the recently created commonwealth environmental water holder will manage the water rights that the Commonwealth is currently buying.
- **Right to return flow**
  In the US and Chile, downstream water rights holders have a right to return flow. Water rights holders may only transfer the quantity equivalent to the consumptive use of a water right. The transferable quantity of a water right is usually difficult to determine. Conflicts over water transfers are usually ruled in Civil Courts between private parties. However, the Northern Colorado Water Conservancy District (Big Thompson Project) is the owner of the return flow. In Australia, Mexico and Spain, water rights holders have no right over the return flow.
- **Dispute over water uses**
  In the US and Chile, disputes are usually treated before Civil Courts, while in other states they are resolved by special Water Administrative Courts.

Although water markets are increasingly presented as models for water reform worldwide, there are many obstacles in implementation. In fact, even in the regions where they have been introduced for many years, their implementation remains very weak. Water markets are usually not very active. The principal difficulties to a well-functioning water market is a lack of infrastructure facilitating water transfers, problems in measuring and controlling water diversions, and a sometimes insufficient number of water rights holders. Overall, these difficulties are linked to those characteristics of water resources which can only be subject to market systems in very particular conditions in a specific, limited area. In any case, water markets can only be a limited solution to water management. Water markets, because they only address the stock side of water resources, will always be insufficient. A comprehensive water plan is necessary to take into account the other aspects of water resource management. As we have tried to show, there is significant variation in the level of regulation between water markets. Institutional frameworks can vary substantially depending on the definition of the water rights, the transfer procedures, the rules in case of shortage, the provision for the environment and the resolution of water rights disputes.

CONCLUSION

The recognition of water as an economic good through the introduction of market systems faces many challenges. These challenges are not only technical and geographical but also institutional. To obtain equitable, sustainable and efficient outcomes, water markets should be integrated into a more general water planning. The institutional framework surrounding water markets is essential to allow the resource to be treated in all its aspects: economic, social and environmental. The property regimes governing water resources are revealed by these different decision-making procedures and institutional frameworks.

Although many countries have introduced important water reforms, their implementation is weak in most places. In many respects, it seems that we are still in the learning phase of how to design and implement policies recognizing the economic dimension of water resources. The implementation of these policies is rendered difficult by the natural characteristics of water resources. Water resources and water uses are not easily measurable and controllable. Water resource economics is definitely a field that will develop in the future, but economists will not be able to contribute to the improvement of water resource management if they are not associated with hydrologists and natural scientists able to understand the complex interconnection between the different water uses and the larger environment. Water entitlements and pricing policies necessitate integrated management to ensure all types of water resources are regulated. This means that these policies require, for example, a very good knowledge of underground water resources and their levels of renewability. Because the policies recognizing the economic dimension of water resources necessitate integrated management, they will sooner or later concern international watercourses. In other words, the trend that one observes at the domestic level in many countries from riparianism towards an administrative system of water entitlements will probably touch international watercourses. We tried to make clear, however, that in domestic politics the recognition of water as an economic good has usually been accompanied by the inclusion of water resources into the public domain of the states. At the international level, it is argued here that the recognition of water as an economic good will require the definition of international watercourses as common property. The next part sheds light on the lack of clear definition of the property regime of international watercourses. And this is demonstrated by the fact that international trade rules increasingly interfere with international water law.

PART II

Trade in Water and International Water Law

Of Property and Interests

International water transfers have recently given rise to important controversies. Although few of these projects have yet been implemented, they raise fundamental questions regarding the property regime of water resources, i.e. namely, to whom does water belong? This question is even more controversial when the transfer concerns an international water system such as the Great Lakes or the Rhône River. What are the rights of non-riparian countries to international watercourses? To what extent are water exports toward a non-riparian state an 'equitable and reasonable use'?

Recently, in summer 2008 the transfer of Rhône River water to Barcelona, Spain by tanker occurred in a legal vacuum. As the quantities of water involved in the sale were relatively small and the transfer happened on a very short term, the controversy did not really trigger huge reactions. Nevertheless, such transactions clearly show the absence of rules governing international water transfers from international watercourses.

It is argued here that international water law is not yet well equipped to face these new challenges. Or, more precisely, in most places, international watercourses are not governed by a well-functioning, common decision-making body able to enforce clear regulations on those matters. Because there are no clear

ISBN: 9781843393610. Published by IWA Publishing, London, UK.

rules to address these new developments, international trade law is increasingly used to fill the legal gaps.

What are the principles governing international watercourses today? How do they contribute to the definition of the property regimes of international watercourses? Why and how does international trade law interfere with the law of non-navigational uses of international watercourses? The following part of our book is devoted to answering these questions.

Chapter 4

The Property Regime of International Watercourses

Domestic water reforms recognizing water as an economic good are usually accompanied by the definition of water resources as common property and the inclusion of water resources in the public domain. This has been done in various ways with various implications. Thus, water resources have been declared 'state property', 'public property' or the 'common heritage of the nation'. At the international level however, it has been much more difficult to define the property regime of international watercourses. The following part reviews the existing rules governing international watercourses in an attempt to identify the property regime of international watercourses.

DOES INTERNATIONAL WATER LAW EXIST?

This is a particularly provocative question for international experts in international water law, for whom the issue does not even come to mind. To raise this question is, however, a good way to present the complexities and challenges of the field. It is sometimes argued that because there is no supra-national authority and a general lack of enforcement mechanisms, international water law does not

 International Trade in Water Rights: The Next Step. By Aline Baillat. ISBN: 9781843393610. Published by IWA Publishing, London, UK.

exist. This is, however, an incorrect understanding of the nature of public international law. Public international law and therefore international water law apply to sovereign states and it is primarily up to states themselves to ensure compliance with international commitments. A second set of arguments – more appropriate – concerns the identification and content of the applicable rules. For some, the absence of a sufficient number of ratifications more than 10 years after the adoption by the UN General Assembly of the 1997 UN Convention on the Law of Non-navigational Uses of International Watercourses, and thus the fact that it has not yet entered into force[1], is an illustration of the lack of consensus on the content of the core principles of international water law. Finally, the vagueness and ambiguity of the basic principles of international water law are denounced as significant obstacles to their implementation.

Apart from the rules applying to navigation, it is true that the development of a particular law for the use of international watercourses is a relatively new phenomenon. These rules were introduced in the 20th century as a result of economic development and industrialization, including the introduction of hydropower, growing water withdrawals for agricultural, industrial and urban uses, as well as pollution. International water law might be relatively new, but it remains, nevertheless, a true body of international law.

In an important case, the Gabcikovo-Nagymaros Case, the International Court of Justice referred explicitly to the 1997 UN Convention thus granting it international recognition[2]. It is important to recall that international law is found not only in treaties but also in international custom, in the general principles of law and the writings of 'highly qualified publicists' (Article 38 Statute of the ICJ).

The 'equitable and reasonable utilization' principle and the 'obligation not to cause significant harm' are widely recognized as the two cornerstone principles of customary international water law and have been codified in the 1997 UN Convention on the Law of Non-navigational Uses of International Watercourses.

---

[1] For the Convention to enter into force, it is required that at least 35 states ratify, accept, approve or access to the Convention (article 36 of the Convention). There is no time limit for that. As of 2009, there were 18 parties to the Convention (Finland, Germany, Hungary, Iraq, Jordan, Lebanon, Libya, Namibia, The Netherlands, Norway, Portugal, Qatar, South Africa, Sweden, Syria, Uzbekistan and Tunisia and Germany in 2009). Sixteen states signed the Convention (the Convention was open to signature from 21 May 1997 until 21 May 2000). Some states signed the Convention but have not yet ratified, accepted, or approved the Convention: signatories but not yet parties: Côte d'Ivoire, Luxembourg, Paraguay, Venezuela, Yemen (http://untreaty.un.org/). There is no need to be a signatory to become a party to the Convention.

[2] Case concerning the Gabcikovo-Nagymaros Project (Hungary/Slovakia) ICJ. September 1997

The Implicit Common Property Regime of International Watercourses in the 1997 UN Convention on the Law of Non-navigational Uses of International Watercourses

The Convention defines a 'watercourse' as a 'system of surface waters and groundwaters constituting by virtue of their physical relationship a unitary whole and normally flowing into a common terminus'[3]. An international watercourse is a 'watercourse, parts of which are situated in different states' (article 2).

Looking at the text of the UN Convention, one notices that it only addresses questions of 'uses' and never explicitly mentions the property regime of international watercourses. There is neither reference to 'property' nor to 'ownership' of these rivers. Article 1 states clearly that 'the present Convention applies to *uses* of international watercourses and of their waters for purposes other than navigation'. A brief parallel could be drawn with the 1982 UN Convention on the Law of the Sea (UNCLOS) where the legal statuses of the territorial sea, the waters forming straits used for international navigation, archipelagic waters and the 'Area' are clearly defined. Notably, the preamble of the Convention and article 136 explicitly defines the legal status of the 'Area': 'the Area of the seabed and ocean floor and the subsoil thereof, beyond the limits of national jurisdiction, as well as its resources, are the common heritage of mankind'. Article 137 specifies further 'No state shall claim or exercise sovereignty or sovereign rights over any part of the Area or its resources, nor shall any state or natural or juridical person appropriate any part thereof. No such claim or exercise of sovereignty or sovereign rights nor such appropriation shall be recognized'.

Contrasting with the UNCLOS, which clearly defines the legal status of the different zones of the sea, the UN Convention of the Law of Non-navigational Uses of International Watercourses does not mention explicitly the legal status of international watercourses. Ruling uses only, the drafters get around the thorny issue of the property regime of international watercourses.

But looking closely at the spirit and intention of the Convention, one could argue that an international watercourse belongs to all the riparian states. An international watercourse is a common property resource. Indeed the Convention clearly mentions that each state whose territory is part of an international watercourse

[3] In December 2009, the UN General Assembly adopted a resolution on the Law of Transboundary Aquifers (A/RES/63/124). These draft articles apply to all types of transboundary aquifers, whereas the 1997 UN convention applies to groundwaters only when they are related to surface waters. Despite the fact that the general scope of the 2009 draft articles is somewhat broader than in the 1997 UN convention (cf article 1), our analysis remains relevant. The property regime of international surface and underground water resources is implicit in these documents.

has a right to take part in the negotiation over the uses of the watercourse (articles 4-1 and 4-2)[4]. As mentioned above, the decision-making procedure is the most direct illustration of the property regime. Further, article 8 recalls that 'sovereign equality' is a core principle governing the relationship among riparian states along with 'territorial integrity'[5]. In other parts of the Convention, the protection of the international watercourse is presented as a shared responsibility. The riparian states have the 'duty to cooperate in the protection' of the watercourse (article 5-2). Thus, one finds many indications of a property regime of international watercourses. However, this property regime is implicit.

Even though the Convention is a 'framework' convention, whose task is essentially to define broad guiding principles, it could have explicitly addressed the question of the property regime of international watercourses as a common resource shared by riparian states. The absence of such a mention is the result of disagreements among negotiating parties.

The same kind of difficulties emerged during the debate over the adoption of the draft articles on the law of transboundary aquifers. The use of the term 'transboundary' instead of 'international' in the 2009 draft articles may even be interpreted as a step back. Interestingly, Eckstein reports that during the UN International Law Commission's work the fact that groundwater resources were initially addressed within the rubric of 'shared natural resources' raised strong reactions. Some states were opposed to the use of the term 'shared' because they feared this would imply a collective ownership of the resource. Referring to Yamada's records of the debates (special rapporteur), he explains: 'The term 'shared' intimates collective ownership and suggests that the resource at issue may be subject to common or equal ownership and potentially to the common heritage of humankind' (Eckstein 2007, p553).

The legal status and property regime of international watercourses and aquifers are thus to be determined on a case by case basis by the riparian/aquifer states in each international water system. Negotiations over such issues will be triggered sooner or later by domestic water reforms and the clarification of water rights within countries.

---

[4] Art 4-1: 'every watercourse state is entitled to participate in the negotiation of and to become a party to any watercourse agreement that applies to the entire international watercourse, as well as to participate in any relevant consultations.' Art 4-2: 'a watercourse state whose use of an international watercourse may be affected to a significant extent by the implementation of a proposed watercourse agreement that applies only to a part of the watercourse or to a particular project, programme or use is entitled to participate in consultations on such an agreement . . . .'

[5] Article 8: 'Watercourse states shall cooperate on the basis of sovereign equality, territorial integrity, mutual benefit and good faith in order to attain optimal utilization and adequate protection of an international watercourse'.

The Evolution of the Property Regime of International Watercourses

International water law is continually evolving. Before the gradual recognition of the equitable and reasonable use doctrine in customary international water law, there were two different, conflicting doctrines. On the one hand, upstream states defended their interests by putting forward the 'absolute territorial sovereignty' doctrine, arguing that they are entirely sovereign to undertake any kind of work on the part of the watercourse situated on their territory. This doctrine is also associated with the Harmon doctrine. In a dispute over the Rio Grande, US Attorney General Judson Harmon replied in 1895 to Mexico's protest and formulated its famous 'doctrine':

> 'The fundamental principle of international law is the absolute sovereignty of every nation, as against all others, within its own territory [. . .]
>
> All exceptions [. . .] to the full and complete power of a nation within its own territories must be traced up to the consent of the nation itself. They can flow from no other legitimate source' (quoted in MacCaffrey 2001, 115).

On the other hand, downstream states used a version of the natural flow doctrine of common law to block the development of any types of work upstream. This doctrine is the opposite extreme position to the absolute territorial sovereignty; it is called the 'absolute territorial integrity'. Under this doctrine, the upstream state may do nothing that might affect the natural flow of the waters into the downstream state.

As we will see later, these two opposing views progressively evolved toward the doctrine of limited territorial sovereignty expressed in the core principle of equitable and reasonable use.

Some authors would prefer that international water law moves beyond the principle of equitable and reasonable use toward the doctrine of community of interest or common management. According to this doctrine, the riparian states of the same watercourse form a community tied together by the physical unity of the shared watercourse. This doctrine exists in two versions. The soft version considers that it is the *interests* in the watercourse that are common.

The 1929 decision of the Permanent Court of International Justice on the River Oder is famous for the introduction of the principle of 'community of interest of riparian states'. It states

> 'this community of interest in a navigable river becomes the basis of a common legal right, the essential features of which are the perfect equality of all riparian states in the use of the whole course of the river and the exclusion of any preferential

privileges of any riparian state in relation to others'[6] (the case dealt with a question on navigational uses).

The strong version considers that the watercourse is a common *property*. Grotius in *De Juri Belli Ac Pactis* wrote:

> 'thus a river, viewed as a stream, is the property of the people through whose territory it flows, or of the ruler under whose sway that people is [. . .] the same river, viewed as running water, has remained common property, so that any one may drink or draw water from it' (quoted in MacCaffrey 2001, 150).

Today, the recognition of international watercourses as common property seems to still be far away. However, some regional instruments such as the 1995 Protocol on shared watercourse systems in the Southern African Development Community (SADC) region recognized the concept of 'community of interest' (article 2 .2.). Also, the integration of the environmental protections requirement and the multiplication of references to the river basin as a unit of management in international instruments are an illustration that international water law has started to evolve in the direction of the concept of 'community of interest'.

Authors like MacCaffrey or Sylvie Paquerot propose an even more progressive theory. MacCaffrey explains:

> 'the concept of community management can be taken further [. . .] the time has come to view freshwater in global terms. Uneven distribution of the world's freshwater means that some states, and regions, have more than they need, while others have less [. . .] The concept of 'equity' and 'equitable allocation' should be applied to this problem. It will not be easy. But it seems undeniable that the international community has a vital interest in the global hydrological cycle, just as it has a vital interest in other great natural systems, such as the Earth's climate' (MacCaffrey 2001, 172).

For Sylvie Paquerot, freshwater (not only water in international watercourses) should be recognized as the common heritage of mankind. According to her, we should go beyond the logic of common property to include not only local public interest but also global public interest. The implementation of such a concept would require the creation of a world water organization.

Even though these last proposals are less realistic, they do have some merit, in part because they raise the question of non-riparian states' rights to an international watercourse. This question has recently been highlighted by different projects of international water transfers. In addition, reflecting on the recognition of water as a human right and its implication for international watercourses management, some

[6] Territorial Jurisdiction of the International Commission of the Oder River [1929], P.C.I.J. (Ser.A) No. 23, (Sept. 10).

authors (MacCaffrey 1992) have interpreted the human right to water in a way that justifies the right of non-riparian states to international watercourses' water. Thus, international solidarity can be considered a motivation for international water transfers from international watercourses.

Grey Zones in International Water Law

In the 1997 UN Convention codifying customary international water law, the property regime of international watercourses is implicit at best. In fact, there is still disagreement over which type of property regimes should govern international watercourses. These disagreements over the property regime of international watercourses are revealed in the ambiguities and vagueness of the rules and principles of international water law.

The principles of 'equitable and reasonable utilization' and 'no harm' are internationally recognized as principles of law. But many authors denounced the vagueness of these principles as the reason why international water law is so poorly implemented. For others, however, the flexibility of this principle is precisely its strength (Wouters 2001).

Beyond the question of the vagueness -or flexibility- of the 'equitable and reasonable utilization' principle, is the debate over the hierarchy between the principle of 'equitable and reasonable utilization' and the 'no harm' rule. Carefully examining the negotiations preceding the General Assembly adoption of the 1997 UN Convention on the Law of Non-navigational Uses, Dellapenna identifies the key disagreements and difficulties in customary international water law. During the negotiations some states opposed the subordination of the 'equitable and reasonable utilization' to the 'no harm' rule arguing that a strict application of this rule would prohibit any new development by upper-riparian state. Dellapenna explains:

> 'In practical application, the no harm rule resembles the natural flow theory of riparian rights in US law and also the rule of prior appropriation as found in the western states of the USA [. . .] To treat priority in time as controlling, or even dominant, would replace the balancing of need and interest characteristic of equitable utilisation with an absolute rule derived from history rather than from geography' (Dellapenna 2001, 280).

For him, the rewriting of article 7 - Obligation not to cause significant harm – is a clear manifestation of the willingness of some delegations to give the primacy to the 'equitable and reasonable utilization' rule over the 'no harm' rule[7].

---

[7] 'As if this were not enough to justify a conclusion that the UN Convention makes the obligation to prevent harm subordinate to the rule of equitable utilisation, this relationship is made explicit in paragraph 2 of the article where the obligation to take 'appropriate measures', as well as the obligation

According to MacCaffrey, on the contrary, although the texts have changed, the no harm rule remains primary over the rule of 'equitable and reasonable utilization' (MacCaffrey 2001, 254). However, as he further explains, 'Perhaps not surprisingly, the compromising formula arrived at in the UN negotiations is a *pot-pourri* containing something for everyone' (Ibid., 255).

Linked to the ambiguity over the relationship between the two core principles of international water law is the important but unanswered question of priority among uses. In the 1997 UN Convention on the Law of Non-navigational Uses, the absence of clear rules over priority of uses -except for 'vital human needs'[8]- is perceived by some international water law specialists as an opportunity for more flexible and adaptable solutions to the particular circumstances in each watercourse (Wouters 2001).

Some other important ambiguous zones in international customary law of water resources can be identified by comparing different codifications of international water law, such as for example the 1966 Helsinki Rules, the 1997 UN Convention but also the more recent 2004 Berlin Rules.

In international water law, the question of compensation is not yet clarified. The 1997 UN Convention on the Law of Non-navigational Uses makes provisions for compensation in the case of 'significant damages' only. The Helsinki Rules, however, mention compensations as 'a means of adjusting conflicts among uses and during negotiations' (article 5.10). The extent to which a state can ask compensation – including cash compensation - in exchange for concession over water uses remains unclear. Is a state entitled to ask compensation from other basin state using its unused allocation? Article 15[9] of the Berlin Rules makes no reference to compensation in the case where a basin state uses part of the allocation of another basin state. The commentary of the

---

to 'discuss compensation', are to be made with 'due regard to the provisions of article 5 and 6' – the principle, in other words, of equitable utilisation.' (Dellapenna 2001, 285)

[8] Article 10 UN Convention: Relationship between different kinds of uses

1. In the absence of agreement or custom to the contrary, no use of an international watercourse enjoys inherent priority over other uses.

2. In the event of a conflict between uses of an international watercourse, it shall be resolved with reference to articles 5 to 7, with special regard being given to the requirements of vital human needs.

[9] Article 15: Using Allocated Water in Other Basin States:

1- Allocation by agreement or otherwise to one basin state does not prevent use by another basin state to the extent that the basin state to which the water is allocated does not in fact use of the water.

2- Use of a water for purposes of this Article includes water necessary to assure ecological flows or otherwise to maintain ecological integrity or to minimize environmental harm.

3- Use of water by a basin state other than the one to which the water is allocated does not preclude the basin state to which the water is allocated from using the water when it chooses to do so.

article stressed that: 'the question at all times is whether the use in question is equitable and reasonable'.

Linked to this important question is the issue of 'future needs'. The 1997 UN Convention remains silent over future needs – probably because no agreement could be reached over them. The Helsinki Rules, however, clearly provide that 'a basin state may not be denied the present reasonable use of the waters of an international drainage basin to reserve for a co-basin state *a future use* of such waters'(article 7). There are two ways to interpret this rule. In a certain manner, this rule may appear inequitable and disadvantageous for developing countries. This rule seems to privilege developed over developing countries by giving priority to existing uses over future uses and needs. From another point of view, one notes that this rule offers a guarantee against thinking in terms of 'ownership' of river waters. Riparian states have an equal right to use the river but this does not correspond to an equal share of the water, nor does it correspond to an ownership right over a certain part of the river. In other words, water rights cannot be considered apart from specific uses, they exist only in relation to present uses or projects. International watercourses cannot be appropriated. This interpretation is confirmed by the more recent Berlin Rules in the commentary of article 15.

The uncertain hierarchy between the 'equitable and reasonable use' principle and the 'no harm' rule, the lack of priority among uses, the question of compensation for concessions over unused allocation and the rights over future needs form 'the grey area' of customary international water law. The 2004 Berlin Rules offer however significant improvements in the codification and development of customary international water law by including emerging elements of environmental, humanitarian and human rights international customary law. Another major evolution is the application of these principles not only to international waters but also to national waters. Important questions are however still not clearly answered. International watercourses are not governed by a clear property regime. In international water law, the common property regime of international watercourses is implicit at best.

Current State Practice over International Watercourses

After reviewing the main principles of international water law – including their ambiguities - it is important to turn to actual state practice.

Some optimistic studies offer a positive view of international water law recalling for example successful agreements like those over the Indus or the Mekong Rivers despite conflictual settings. Recent agreements such as the South African Development Community Protocol on Shared Watercourses are also presented as the significant expression of regional cooperation in international river basins. In addition, the successful contribution of the World Bank to the implementation of international water law is often put forward. The World Bank, along with other multilateral donor agencies, committed themselves to not finance any projects on international watercourses before an agreement is reached among all riparian states and environmental impacts are carefully assessed. Nonetheless, controversial projects such as the Three Gorges Dam in China, the Narmada Dam in India and the GAP in Turkey have been undertaken without - or almost without - the aid of the international community. Offering an opposite view on international water law, Waterbury mentions, 'most international river basins are not governed by any comprehensive agreements on the use and allocation of the basin's water' (Waterbury 1997, 279).

Aaron Wolf and his colleagues, working for the Transboundary Freshwater Dispute Database project at Oregon State University, attempted to draw a picture of current state practice over international watercourses. Their treaty database includes 145 treaties concerning non-navigational uses of international watercourses (http://www.transboundarywaters.orst.edu/). They drew out from these treaties general trends in negotiations over international watercourses. State practice is not limited to treaties but they are nonetheless a good indication of general patterns in negotiations over international watercourses (some treaties indeed will remain a dead letter forever, whereas successful agreed practices may never be formalized).

A first interesting conclusion is that almost all the treaties studied are bilateral treaties (124 out of 145 treaties) – whereas the 263 international basins of the world are generally shared by more than two states. This is a clear indication that the principle of sovereign equality of all riparian states is not enforced and that the principle of community of interests is not yet a reality in most of the world's river basins. This is also an illustration that the concept of 'integrated river basin management' is still far from being implemented.

Most treaties focus on hydropower and water supplies, industrial uses and pollution. But only a few treaties clearly allocate water. Wolf summarizes by noting that 'the 145 treaties which govern the world's international watersheds, and the international law on which they are based, are in their respective infancies. More than half of these treaties include no monitoring provisions whatsoever and, perhaps as a consequence, two-thirds do not delineate specific allocations and four-fifths have no enforcement mechanism' (Hammer and Wolf 1998).

It is extremely difficult to provide a good description of current practices in the management of international watercourses. This field, like the rest of international relations, is shaped by diverging views on human nature and world order. The reality is probably somewhere in-between these optimistic and pessimistic world understandings, in a particularly complex form of interdependence mixing both conflict and cooperation. Besides, it is difficult to evaluate international water law as it is implemented very differently from one region to another.

In fact, the lack of definition and enforcement of property regimes has not yet been crucial for the management of international watercourses. For the time being, the principle of 'equitable and reasonable utilization' has been implemented in various ways and in various circumstances without any big difficulty in most regions of the world.

For Dellapenna, international water law principles are suitable only as long as water uses are not too conflicting. He argues that international water law principles are analogous to the riparian rights system usually used in well-watered regions. For him, however, in regions where water is scarce, international water law principles do not offer useful guidance.

Highlighting some of the weaknesses of customary international water law, Dellapenna mentions:

> 'to begin to examine why the customary international regime fails first consider the experience of the USA. [. . .] The situation would be even worse if the actors were to measure the right to use water by a vaguely defined equitable apportionment. The closest analogue to this system is the riparian rights system (and its interstate analogue of 'equitable apportionment') as applied in the eastern USA. [. . .] The 'reasonable use' theory of riparian rights has barely functioned in areas of the USA that are without chronic water shortages and that have a strong judicial structure to resolve disputes between users. Whenever water use in the eastern USA outstrips the available sources of water, riparian rights have been abandoned in favour of a new system of water rights that are heavily administered by state agencies that allocate water to particular uses by time-limited permits and leave authority in the agencies to determine the most socially beneficial ('reasonable') use of the water' (Dellapenna 2001, 275-276).

Dellapenna concludes from the US experience that the principles of customary international water law are too vague and too controversial to provide useful guidelines to define clear water rights- especially in regions where water demand exceeds water supply. He seems to suggest the evolution of international water law towards a system of administrative water rights. According to him, the principle of equitable and reasonable use in international water law is to follow the same path as the principle of riparian rights in domestic water law, now replaced by clearer systems of water rights.

In our view, such an evolution towards the introduction of administrative water rights systems along international watercourses will be possible and desirable only if international watercourses are first defined as common property. The definition and implementation of the common property is a necessary preliminary step before the introduction of international water rights systems.

Dellapenna's analogy between the domestic and international level is interesting but lacks an important element: the definition of the property regime of water resources. Indeed, at the domestic level, water resources are more easily defined as 'state property' or 'common property' than at the international level. As we saw in the first part, such a definition of the property regime usually preceded or accompanied the clarification of water rights and water allocation systems in domestic water reform.

In any case, Dellapenna's analogy between domestic and international water law is useful. We would like indeed to underline that international water law comes from domestic water law and that it will probably continue to evolve in the same direction. This is why the third part of this book is devoted to the study of current federal states' practice (Australia and the USA).

Federal States' Practice as Origin of International Water Law

Federal states' practice and federal court decisions over interstate water allocation played an important role in the development of international water law principles (ILC 1994, 40). The basic principles of 'reasonable and equitable utilisation' and 'no significant harm' found their origins in federal court decisions, particularly those of Switzerland, Germany, and the USA (Kaya 2003, 73; MacCaffrey 2001, 221; Spiegel 2005). On his side, MacCaffrey wrote:

> 'The law of shared water resources has perhaps been developed in the greatest detail by national courts, especially the US Supreme Court in its decisions in interstate apportionment cases. These national court decisions have historically contributed to the development of international law in the field and will most likely continue to do so' (MacCaffrey 2001, 231).

The federal courts mentioned above explicitly stated that a dispute between two sovereign units of their federation had to be resolved by international law principles. Thus, these cases are clear cases of co-construction between the international and national levels: applying international law to federal state circumstances, these courts developed principles suitable for true international disputes.

For example, Hebert Smith, one of the early international water law specialists, referred in his 1931 book to the Swiss *Bundesgericht's* decision of 1878 in a case opposing the cantons of Aargau to Zurich (Jona River). He wrote, 'the case has some interest as being the earliest one, so far as I have been able to discover, in which a judicial tribunal has approached the problem from the standpoint of international law' (Smith 1931, 39).

Indeed, the Court stated:

> 'Especially where the relations among component units of a federal state are at stake, a rule of international law derived from the law of good neighbourliness applies. According to that rule, the exercise of a right may not affect the right of a neighbour. The two rights are equal, and, in the event of a conflict, a *reasonable* arrangement has to be found on the basis of the relevant circumstances' (quoted in MacCaffrey 2001, 228).

The Court further stated that 'the cantons have no private ownership, but only sovereignty' and 'by virtue of its sovereignty each canton has the right to adopt the necessary measures for such *reasonable use* of public waters as corresponds to its needs, provided [. . .] that the common usage of water is not rendered impossible but on the contrary is left open to the other cantons' (Smith 1931, 105).

In the same manner, applying international law principles to a dispute between federal units, the German *Staatsgerichtshof* laid down the basis for the development of the equitable and reasonable use principle and the obligation to prevent significant harm. It also offers solutions for the combination of the two principles. In the Donauversinkung case opposing Baden and Württemburg in 1927, the German federal constitutional court stated: 'the interests of the states in question must be weighed in an *equitable manner* against one another. One must consider not only the absolute injury caused to the neighbouring state, but also the relation of the advantage gained by one to the injury caused by the other'(ILC 1994, 53).

It is generally acknowledged that the principle of equitable apportionment found its origins in the United States. At the beginning of the 20th century, the US experienced a series of important interstate disputes over water which pushed the Supreme Court to develop the doctrine of equitable apportionment. In the key case *Kansas v. Colorado* in 1907, the US Supreme Court explicitly mentioned that principles of international law were relevant to decide a case involving two US sovereign states. Indeed,

> 'Nor in our jurisdiction ousted, even if, because Kansas and Colorado are states sovereign and independent in local matters, the relations between them depend in any respect upon principles of international law. International law is no alien in this tribunal.'

Further the Court stated:

> 'One cardinal rule, underlying all the relations of the states to each other, is that of equality of right. Each state stands on the same level with all the rest.[. .] Yet, whenever [. . .], the action of one state reaches through the agency of natural laws into the territory of another state, the question of the extent and limitations of the

> rights of the two states becomes a matter of judiciable dispute between them, and this court is called upon to settle that dispute in such a way as will recognize the equal rights of both and at the same time establish justice between them' (*Kansas v. Colorado*).

Patricia Jones (2001) mentions the different economic and scientific factors used by the Supreme Court to define equitable apportionment. These include: the degree of dependence of the population on the water (*Kansas v. Colorado* 1907); the benefits for the diverter compared to the harm of the injured states; financially practicable conservation measures (*Wyoming v. Colorado* 1922; *Colorado v. New Mexico* 1982). The scientific factors consist of physical and climatic conditions, consumptive uses by section of the watercourse, character of return flows, and availability of storage water (*Nebraska v. Wyoming* 1945) (Jones 2001).

The principle of 'equitable and reasonable utilisation' codified in the 1997 UN Convention is very close to the Supreme Court's doctrine of 'equitable apportionment'. Article 6 of the UN Convention indeed mentions various factors relevant to the definition of 'equitable and reasonable utilisation' which are very similar to the US Supreme Court's scientific and economic criteria: geographic, hydrographical, hydrological, climatic, ecological and other factors of a natural character; social and economic needs; the population dependent on the watercourse; the effects of uses on other watercourse states; existing and potential uses of the watercourse; conservation, protection, development and economy of use and the costs of measures taken to that effect; the availability of alternatives, of comparable value, to a particular planned or existing use.

The 1931 Supreme Court decision in *Connecticut v. Massachusetts* was important for the development of the distinction between the equality of rights and the equity principle in international water law. The Supreme Court states, 'As was shown in *Kansas v. Colorado* . . .such disputes are to be settled on the basis of equality of right. But this is not to say that there must be an equal division of the waters. . . .' (Kaya 2003, 90). The International Law Commission in its commentary to draft article 5 of the UN Convention on International Watercourses adopted the same kind of reasoning as the Supreme Court to clarify the relationship between the principles of 'equality of right' and 'equitable and reasonable utilization'. The commentary mentions that indeed 'this fundamental principle of 'equality of right' does not, however, mean that each watercourse state is entitled to an equal share of the uses and benefits of the watercourse. Nor does it mean that the water itself is divided into identical portions. Rather, each watercourse state is entitled to use and benefit from the watercourse in an equitable manner. The scope of a state's rights of equitable utilization depends on the facts and circumstances of each individual case, and specifically on a weighing of all relevant factors, as provided in article 6' (ILC 1994, 29).

Federal state practice played an important role in the development of international water law principles and will probably continue to do so. Australia and the USA have recently introduced new tools for the management of their watercourses shared by their different federal units. These recent developments are worth being studied as the potential evolution of the management of truly international river basins. This will be done in the third part of this book.

INTERNATIONAL WATER TRANSFERS AND INTERNATIONAL WATER LAW

The limits of international water law and, more specifically, the lack of definition in property regimes of international watercourses, have been brought to light by various projects of international water transfers.

Water transfers, which are very much emotional and controversial projects, are being increasingly studied (Lasserre 2006). In these studies, the technical and political aspects of such projects are usually addressed along with the consequences of the application of international trade law to water resources. The true problem with these transfers is the lack of a property regime governing international watercourses, not the question of whether or not international trade law should be applicable. As can be gleamed from the case studies, the distinction between trade in water and trade in water rights is often neglected.

An international water transfer is the systematic removal of bulk water from one country and transferred to another. There are different modes of transfer: by pipeline, by supertankers or by water bags towed behind barges[10]. These transfers may be intra- or inter-basin depending on whether they take place in the same river basin or between different river basins. International water transfers may also be undertaken by different actors: they either take the form of international treaties between two states or international commercial contracts involving private companies, municipalities, etc. The term 'international water transfers' is to be distinguished from the phrase 'water exports' - 'water exports' referring to a particular type of water transfer in which bulk water is considered an international tradable commodity.

In fact, water is already being transferred across national borders, but essentially on a case-by-case basis to face urgent water crises (like Japan, Israel, Nauru,

[10] Spragg bags: http://www.waterbag.com, last visited in May 2010.

Cyprus, Spain etc.). These transfers by tankers are very costly and are usually not considered a long-term solution (Gleick 2002, 45). However, proposals of large-scale and long-term water transfers across international borders are increasingly being considered in many parts of the world (essentially through pipelines). The water transfer between Lesotho and South Africa, in place since 1998, is considered to be the first ever international water transfer. A project of an underwater pipeline between Turkey and Israel to transport water and other resources is currently under feasibility studies. Another important pipeline project between France and Spain involving the Rhône River recently remerged as a consequence of an ongoing drought in Catalonia. It is difficult to figure out if such international water transfers will really develop in the future or not. Peter Gleick concludes his analysis of this phenomenon by saying that: 'our analysis suggests that large-scale, long-term bulk exports of water across international borders are unlikely for many reasons, especially the high economic cost of moving water' (Gleick 2002, 54).

If one considers the increasing number of large-scale water transfers at the national level, it seems reasonable to take seriously the option of the development of international water transfers (i.e. in Libya, Western USA, Egypt, China, India, Spain, etc.) As an example, the Waterbank.com Company proposes more than 250 sources of bulk water for sale on its website (New Zealand, Alaska, Eastern US, Greenland, Sweden, etc.). For the time being, however, only a few international trade operations have occurred. Will this form of commerce develop? The trade of water by supertanker is supposed to remain limited in scale and time because of the high economic cost of transport. But a recent agreement between Dominica and a US company suggests that, in some cases, water transfers by tankers might be economically viable. Sisserou Water Inc., a Colorado-based company, explains in its website: 'Our unique shipping system ensures delivery of Sisserou Water around the world at competitive prices for an unparalleled product. Healthier and cheaper that desalination, our water product provides the viable solution for many water starved regions of the world'[11]. In November 2009 the Commonwealth of Dominica signed a licence in favor of Sisserou Water Inc. to extract 11 million m^3 of water per annum for export (Commonwealth of Dominica 2009). Although the case involves bottled water, it is worth mentioning because it shows that water is increasingly moved over long distance. Sisserou Water Inc. expects to export Dominican water to some US states but also potentially to Qatar and Yemen. Another bulk water export project between Madagascar and Saudi Arabia illustrates this new trend of poor countries selling their water to foreign countries. The feasibility study conducted in July 2009 by Saudi Arabia dealt with water

[11] http://sisserouwater.com/, last visited in May 2010.

from the Faraony River in Madagascar. Officially for 'technical reasons' but also aware of the controversy raised by the project, Rajolina, the president of the *Haute Autorité de Transition*, eventually cancelled the project in fall 2009[12].

As already mentioned, an alternative mode of transport would be water bags, but this technology has not yet really given any successful results. The development of middle-scale international water transfers by pipelines is, however, a more tangible evolution in the long-term. In the discussion over international water transfers there remains an important unknown fact: i.e the price of desalination and the future development of this technology. Interestingly, Kuwait seems to not bet on the evolution of desalination technology only, instead deciding to buy water from Iran. According to an agreement signed between the two countries in 2003, Iran is to supply Kuwait with 900,000 m^3/day (10m^3/sec.) from Karkheh Dam, which is located in the southern province of Khuzestan, through a 330 km pipeline for 30 years (*Khaleej Times* December 29, 2003). The project had been suspended because of opposition from the pan-arabist movement in Kuwait (*Iran Daily* May 10, 2005; *Iran Daily* January 20, 2006). But in June 2007, the project seemed to have been renewed (*Iran Daily* June 26, 2007).

In many respects, water transfers are based on outdated logic consisting in increasing water supply at any cost. Today, the trend in water policy is rather to manage and control water demand. The solutions to water crises are also researched in techniques such as desalination and wastewater recycling. Important research is being undertaken in these fields. The cost of desalination has significantly decreased. Less environmentally damaging techniques have been introduced, such as the use of renewable energy sources for desalination. Desalination still faces economic, technical and environmental challenges (Landais 2009). And desalination is not always a possible and sustainable alternative. A more sustainable solution is wastewater recycling. It seems that a lot could be done in the field of wastewater collection, treatment and reuse.

In this regard, Singapore offers a great model of sustainable water management. This small island city has long been dependent on water imports from Johor, Malaysia. Singapore managed to significantly reduce its dependence on Malaysia by developing, among other things, wastewater collection and treatment techniques. When the two bilateral agreements on water import between Malaysia and Singapore will expire in 2011 and 2061, Singapore expects to be self-sufficient. The national water agency, PUB, is even producing bottled water from wastewater! Sludge is further treated and dried. Biogas created by these processes is used to run the dryers. The remaining sludge is used for agriculture.

[12] One should recall that one of the reasons why Rajolina took power in 2009 was precisely its opposition to the sale of land to the Korean firm Daewoo.

Thus, whether international water transfers projects will really develop or not is uncertain, but it is worth studying them as they raise the fundamental questions of the status and property regime of water resources.

Of course, international water transfers do not necessarily concern international watercourses. Turkey and Israel have been negotiating a water transfer for many years. These proposals concerned Turkish national rivers: Ceyhan, Seyhan and the Manavgat Rivers. Although these projects implicate national rivers, they might have some indirect effects on the way international watercourses are considered. The introduction of bulk water sale and bulk water export highlights the necessity to clarify the property regime of international watercourses. This does not mean that such exports should not take place, but rather that the rules surrounding such exports, especially when they concern international watercourses should be clarified.

Turkish Water Export to Israel

Turkey and Israel have been working on a water transfer agreement for many years.The first transfer proposal was made by then-Turkish Prime Minister Turgut Ozal in the 1986 'peace pipeline project'.

A plan was eventually launched in 1992, but was not adopted until 2004 (by a vote of the Cabinet of 13 to 2, with Finance Minister Netanyahu and Minister without portfolio Sheetrit voting against). It should be mentioned that for a long time the project raised vigorous opposition in Israel, particularly among the Sha party. The 2004 agreement provided for the purchase by Israel of 50 million m^3 of Turkey's Manavgat River water each year for 20 years; Water was to be transported by supertankers, but discussions already evoked the possibility of the creation of a pipeline as a long-term solution. The amount of water to be bought accounted for about 3% of Israel's annual fresh water consumption of 1.5 billion m^3. By 2004, Turkey had already completed the installation on the estuary near Antalya, which had a potential to sell 180 million m^3 of water per year. The estuary, near Antalya, is approximately 325 nautical miles from Israel's Ashkelon port (600 km).

In 2004, the price of water was not agreed upon, but Turkey evoked the original price of 23 cents per m^3 to which transport and purification costs were to be added (55–60 euro cents) (*Anadolu News Agency* June 2000). The mode of transport by supertanker proved to be too expensive and was eventually abandoned.

In May 2006, Turkey and Israel negotiated over another plan involving the construction of five underwater pipelines that will transport not only water, but also electricity, natural gas, fibre optics and oil to Israel (*Turkish Weekly* May 12, 2006). The five underwater pipelines would link the new Ceyhan maritime terminal to the Israeli port of Ashkelon. This project emerged as a result of the construction of the Baku-Tblisi-Ceyhan oil pipeline, inaugurated in July 2006.

This pipeline links the Caspian Sea to the eastern Mediterranean going through Azerbaijan, Georgia and Turkey. The oil piped to Turkey is then to be transferred to Israel. Then, from the Israeli port of Ashkelon, the oil will be pumped through the already-existing pipeline to the port of Eilat at the Red Sea, and from there it will be exported by tankers to India and the Far East (*Jerusalem Post* May 11, 2006). In 2008, Israeli and Turkish officials planned to complete the project three years after the finalization of the feasibility studies, which were expected for 2010 (*Turkish Daily News* July 18, 2008).

If these negotiations are successfully concluded, they will create the first-ever international agreement dealing with the systematic sale of bulk water from one country to another.

The link between these negotiations and international water law is not direct. Indeed, the river concerned by the agreement is a 'national' river that flows entirely within Turkey. It contributes, however, to the recognition of water as a commercial good. If national waters can be sold, it follows that international waters can be sold as well. In this case, what rules should be applied to the transfer of water from an international watercourse to a non-riparian state? What are the rights of non-riparian states? Are commercial uses equitable and reasonable uses? Beyond these issues is the question of the property regime of international watercourses.

Bulk Water Export and Upstream States' Reactions

One interpretation of the Turkish interest in turning water into an international commodity is that it is a way to influence negotiations over the Tigris and Euphrates Rivers. By pushing for the evolution of the status of water on the international scene, Turkey wants to influence the property regime of international watercourses. Turkey never recognized the principle of 'international watercourse' and is one of three countries who voted against the adoption of the 1997 Convention at the UN General Assembly (along with China and Burundi). Turkey instead uses the term 'transboundary rivers' (Turkish Foreign Ministry's water department is called 'Bureau of Transboundary Waterways'), implying that the Tigris and Euphrates rivers are purely Turkish property as their catchments are located in Turkey and are at no point 'international' rivers. Abdelazeem Hammad reports that Turkey's former President Suleyman Demirel said, when he was Prime Minister, that, 'since this natural resource springs from our country, we have the absolute right to exploit it in the way we see fit. Syria and Iraq have absolutely no rights in the waters of Turkey' (Hammad 2000).

Following this reasoning, some politicians and scholars even argue that Turkey should be compensated for the water it provides to Syria and Iraq. David Schenker, a US scholar on the Middle East, declared that, 'water has the same significance for Turks as oil does for the Arabs'. He added, 'Turkey wants to sell its water. When taking into consideration the far-from-cordial relations between Turkey and Syria, I can see no reason why Turkey would be willing to give its water for nothing' (Schenker is from the *Washington Institute for Near East Policy*) (*Turkish Daily News* January 7, 2000).

In many respects, Turkey is the spearhead of a movement trying to transform water resources into an international commodity. Other upstream states followed suit.

The negotiations between Nepal and India over water resources are a good illustration of this new trend. During the negotiation over the 1996 Mahakali Treaty, financial compensation for water was hotly debated. Throughout the negotiation, Nepalese officials were very attached to the principle of 'equality of rights' in the management of water resources between the two states and interpreted this principle in a very strict and limited manner – to them, this principle implied that the Mahakali water would be shared on a 50/50 basis. Rejecting the needs-based approach and arguing that the river water should be shared equally, but also aware that Nepal would not use an important part of its share, Nepalese negotiators proposed that India pays royalties for the right to use river water in excess of 50% of the total flow. India refused this proposal and even introduced special terms in the Letter of Exchange to make sure that Nepal would never ask for any kind of compensation for the unutilised portion of its share of the river water[13]. In addition, India made clear that water, in contrast to power, was no 'commercial matter' (Marty 2001, 185).

Kyrgyzstan adopted the same kind of reasoning concerning the Syr Darya water. The Kyrgyz Parliament passed a law in June 2001 that classified water as a commodity, and later, the Kyrgyz Government announced that it was preparing regulations to charge neighbouring states, including Kazakhstan and Uzbekistan, for the water they use.

This new trend in water negotiation seems to have reached the African continent as well. Abdelazeem Hammad reports that 'during a meeting of a technical organization bringing together all countries of the Nile basin, the Ugandans demanded that both Egypt and Sudan compensate it financially for acting as a natural reservoir for the waters of the Nile, since the equatorial lakes from which the White Nile originates lie in its territory. Thus Uganda too sees water as a national resource which should be subject to the conditions of commercial exchange' (Hammad 2000).

[13] Letter of Exchange: Paragraph 3-b: 'It is understood that section 3 of article 3 if the Mahakali Treaty precludes the claim, in any form, by either party on the unutilized portion of the shares of the waters of the Mahakali River of that Party without affecting the provision of the withdrawal of the respective shares of the water of the Mahakali River by each Party under this Treaty.' Quoted in Salman and Uprety, 2002, p 112.

Finally, the Silala River in Bolivia and Chile provides another case of financial compensation negotiations for an international river[14]. In 2009, after years of controversy, Chile and Bolivia seemed to reach agreement on the Silala waters. According to a preliminary agreement signed in May 2009, Chile – which is downstream- accepted to pay Bolivia- upstream- for 50% of the Silala River. The agreement also provides for the launch of a four-year study to determine the flow and use of the stream's water to help calculate how much Chile should pay in the future. This is a major policy change for Chile, which had always refused to pay for the Silala waters, claiming the Silala River is an international watercourse. Bolivia argued on its side that the Silala River is not a natural river. Bolivia explains that the Silala waters originate from 94 springs in its territory that would infiltrate into the Bolivian ground and then flow into the Bolivian territory, were they not artificially diverted into Chile by means of canalizations and human works. In the end of 2009, Bolivia backtracked on the deal demanding compensation for past use as well. The Bolivian Foreign Minister even declared, 'For more than 100 years Chileans took our water without paying a cent, now we want Chile to pay for 100 percent of the Silala's water' (Ore 2009).

The increasing recognition of water as an economic good, combined with the lack of a clear property regime over international watercourses, has made room for some states to proclaim full sovereignty over the water resources located on their territory and to ask royalties in exchange for the water they let go into downstream states. Contrary to international water law principles, and blurring the distinction between national and international watercourses, upstream states increasingly consider that they are indeed exporting – for free - water to downstream neighbours and therefore can ask them for financial compensation (Turkey, Uganda, Nepal, Bolivia, Kyrgyzstan, etc.). Although these arguments should be understood essentially as bargaining tools in upstream/downstream negotiations, they reveal the difficulty of agreeing on the definition of the property regime of international watercourses.

Resistance Against the Commodification of Water Resources: The Lesotho Highlands Water Project (LHWP)

The agreement between the Republic of South Africa and the Kingdom of Lesotho creating the Lesotho Highlands Water Project (LHWP) is considered to be the first-ever international water transfer (Nakayama 2003). Contrary to the Israeli-Turkish agreement, the LHWP project involves two riparian states. The transfer concerns an international river shared by Lesotho and South Africa. The general

[14] In fact, the core of the dispute between Chile and Bolivia is whether the Silala should be considered as an international watercourse or not.

idea is to divert most of the major tributaries of the Senqu River (40%), which naturally flows south out of Lesotho into South Africa (Senqu/Orange River), into Gauteng Province located in the northern part of South Africa. The project thus transfers water from the mountains of the Kingdom of Lesotho to the industrial heart of South Africa, the highly water consumptive Province of Gauteng. Besides the transfer of water, the project includes a hydroelectric component for Lesotho. In fact, the project is still the biggest and the most complex water scheme in Africa.

This ambitious project, consisting of four phases, is supposed to transfer a total of 70m^3/sec to South Africa by 2020. The first phase, which began in 1989, was entirely completed by 2004. It comprises two giant storage dams (Katse and Mohale Dams), a hydropower station (Muela) and more than hundred kilometres of a transfer and delivery tunnel system reaching to the Ash River across the border in South Africa (see map below). Since the completion of the first phase, an average of 30m^3/sec of water is being delivered to South Africa. After delays due to important controversies over Phase I, the implementation of Phase II was officially announced in December 2008. The construction of Polihali Dam[15] at the confluence of Senqu and Khubelu Rivers will start in 2012 and should bring an additional 17 m^3/sec to South Africa by 2018 (Lesotho Today, 2 April 2009).

Does South Africa pay Lesotho for the Highlands Water?

The Trans-Caledon Tunnel Authority, the South African agency in charge of the implementation and management of the project, makes it crystal clear that '*South Africa does not pay Lesotho for the water. Lesotho does receive a financial benefit but for different reasons*' (http://www.tcta.co.za). According to the LHWP Treaty, South Africa indeed pays royalties each month to Lesotho. But, these 'royalties' are a way to share the benefits of the project as explained below.

Costs and Benefits Apportionment according to the LHWP Treaty

The transfer of water from Lesotho to South Africa is regulated by the 1986 LHWP Treaty between the Republic of South Africa and the Kingdom of Lesotho. Article 10 provides that South Africa is responsible for the costs of the water transfer aspects of the project, irrespective of whether the work is located in Lesotho or South Africa (95% of the entire cost of the project), and Lesotho is responsible for the costs related to the generation of hydro-electricity on its territory (estimated at 5% of the cost of the entire project). Concerning the

[15] Phase II of the initial project planned the construction of Mashai Dam on the Senqu River at Mashai. The options for Phase III and IV are the construction of Tsoelike Dam and a dam at Ntoahae, 40 km downstream of Tsoelike Dam on the lower reaches of the Senqu River (but also dams at Malatsi or Lebelo).

Figure 3 Map of the Lesotho Highlands Water Project (Credit to TransCaledon Tunnel Authority).

benefits, article 12 provides that 'South Africa undertakes to share with Lesotho, *by way of royalty payments*, on the basis of 56% on the part of Lesotho and 44 % on the part of South Africa, the net benefit'.

The method to estimate benefits of each party rests basically on a comparison between two projects: one entirely within South Africa and the other necessitating cooperation with Lesotho (LHWP)[16]. The LHWP project was compared to the so-called Orange-Vaal Transfer Scheme (OVTS). This Scheme provided that water would be pumped from the Orange River southwest of Lesotho and transferred via a canal to the Ash River Outfall, from where it would flow to the Vaal Dam.

[16] The part of the treaty dealing with this issue is quite complicated (art 12) and includes other imaginary projects: one corresponding to a water transfer only and the other including the hydropower component but the details of the comparisons are not fully mentioned in the treaty but in what is called the 'Royalty Manual' (3 volumes).

The feasibility studies and cost analyses found that the LHWP was clearly far more cost effective than the OVTS. The operation and maintenance of the LHWP was also a better option, because it relied on gravity instead of pumps. As a result, the royalty payments are comprised of two elements: a fixed and a variable component. The fixed component is calculated on the benefits gained from the investment difference between the two projects. It is payable monthly to Lesotho up to 2045 (a fifty year period from the beginning of payments). The calculation of the variable part is based on the difference in electricity costs attributable to pumping operations and to the difference in the costs attributable to the operation and maintenance of the alternative projects (art 12-10). The variable components are converted into unit rates and the amount paid per month depends on the quantity of water delivered during that month[17]. Royalties should be paid as long as water is delivered: there is no time limit.

In the spirit of the LHWP Treaty, it is not the water itself that is divided but the benefits gained from the common exploitation of the watercourse. It is true that the term 'royalties' introduces confusion over what South Africa really pays Lesotho for. The distinction between paying royalties for water itself and paying royalties to share benefits from a project is, however, fundamental as it involves the very definition of water as a resource.

South African officials seem to be very committed to the task of clarifying the term 'royalties' in the case of the LHWP water transfer in order to avoid any confusion over the status of water. For South African officials, water is not an international commodity. One might even better understand the position of South Africa by considering that South Africa is located downstream of Lesotho. Moreover, one could add that the Senqu/Orange River flows naturally out of Lesotho and reaches South Africa. In this sense, the benefits to South Africa are not merely to receive specified quantities of water but rather to receive 'quantities of water at a particular locality and at a cost substantially lower than the cost of any other scheme which could afford comparable benefits' (Vorster 1987, 112).

The LHWP is relevant to our analysis beyond the fact that it is a water transfer because it is more generally speaking, a good example of upstream/downstream cooperation and benefit-sharing over an international watercourse.

---

[17] The variable royalties amount to 14.24 cents/m^3. The Government of South Africa pays the Government of Lesotho for the royalties through the South African agent, TCTA. The royalties are paid into a Lesotho Highlands Water Reserve Fund (LHWRF) which is administered by the Government of Lesotho. The annual amount of royalties paid is around 100 million Rand (for the fixed part) and 80 million Rands for the variable part for an annual water delivery of around 580 million m^3. (http://www.tcta.co.za, last visited in May 2010).

The international water transfer between Lesotho and South Africa should however be considered an exception due to its very particular geographical characteristics. The water diverted from Lesotho was anyway flowing into South Africa. Instead of flowing into the southern part of South Africa, the water is delivered in the northern part of South Africa.

Conclusion

The fact that international water transfers are increasingly considered an option for water managers reveals the growing water scarcity in many regions of the world. Although this scarcity is almost never irremediable and usually results from bad management, it nevertheless illustrates the growing pressures on water resources. In this context, the need for the definition of international watercourses as common property and the implementation of common decision-making procedures reveals to be pressing. Water is an economic good. Which rules and institutional frameworks will accompany the recognition of water as an economic good in international watercourses? Will water allocation be left to the free market?

An important literature has started to consider the consequences of the application of international trade rules to water resources. For us, however, this literature overlooks the fundamental issue of the property regime of water resources. If trade rules are increasingly applied to water resources it is precisely because there is a legal and institutional vacuum. There is a lack of rules emanating from the river basin states to face the new challenges of the 21st century such as out-of-basin water transfers. They usually have not defined the property regime of their shared watercourse and it has not been implemented into a strong institutional and regulatory framework.

It is true however that the adoption of common decision-making procedures and adaptive management rules is clearly complicated by the international nature of the watercourse. The geographic and hydrologic characteristics of the international watercourse can also complicate the matter. In other words, it might be more difficult to recognize the common property of an international 'successive river' (i.e. where the different jurisdictions are essentially located upstream/downstream) than of an international lake or border river. In the same manner, the fact that groundwaters are invisible complicates the task for those trying to determine the common property regime of an international aquifer. Nonetheless, for example, it is not because the transfer of Rhône water from Marseille to Barcelona does not directly modify the flow in Switzerland that Switzerland should not have a say in such a transfer. The recognition of the common property of the entire Rhône River shared by Switzerland and France would start with the creation of a bi-national river commission (and not just for the Leman Lake). It is true that for the time being apart from the management of the Leman Lake there was no need to create such a bi-national commission, given the absence of scarcity in the region. However with the question of Rhône water transfer to Barcelona, especially if it becomes permanent, it will be necessary to design the 'international layer' of the web of interests of the Rhône River through notably the creation of a bi-national river commission and the adoption of rules governing out-of-basin transfers. Under this scenario, *Agua de Barcelona* would have to ask the bi-national commission for a permit to transfer and use Rhône water[18]. The bi-national commission would then review the proposal according to its own rules and criteria (protection of the area of origin, protection of the ecosystem, requirement of conservation measures, entity allowed to ask for permit, duration, quantities, compatibility with the basin plan, etc.). Whereas trade rules may apply to trade in (unregulated) water, in such a scenario, the rules applicable would be the rules of the water rights systems established by the bi-national river basin commission.

If such frameworks were implemented, the question of international water transfers would not deal with international trade in water but instead with international water rights (potentially tradable). The next chapter is an attempt to demonstrate this through the example of international water transfers projects from the Great Lakes.

[18] In spring 2008, Agua de Barcelona directly negotiated with the *Compagnie des Eaux de Marseille* and the *Société du Canal de Provence*. The French company and society sold the Rhône water as if it were their ownership whereas they only have a concession to use the water. In such a case, there should be another authority in charge of negotiating and approving the transfer.

Chapter 5

The Relationship between International Trade Law and International Water Law

With the emergence of various projects of international water transfers and exports, the question whether water should be subject to international trade rules is raised. Thus, an important literature developed over the legal consequences of the recognition of water as an international tradable commodity. This literature usually highlights that the question should not be limited to regulations concerning trade in goods but should also include investment provisions and potential regulations over trade in services (De Haan 1997, Al-Assar 2002, Matsuoka 2001, Girouard 2003, Cossy 2005; Brown Weiss, Laurence Boisson de Chazournes, and Nathalie Bernasconi-Osterwalder 2005).

The basic question concerning water and international trade in goods is whether water can be qualified as a 'product' under GATT. Whereas bottled water is clearly a product under GATT, the answer remains unclear concerning 'water in its natural state'[1]. A general consensus acknowledges that a 'product' results from a minimum

[1] 'Water in its natural state' is a phrase used in the 1993 gentlemen's agreement among NAFTA parties to exclude 'water in its natural state' from NAFTA provisions. For a discussion of the legal value of such a document, see IJC 2000.

ISBN: 9781843393610. Published by IWA Publishing, London, UK.

human intervention and a transformation process but the question remains over the degree of intervention. When does water become a product?

By entering into international trade agreements such as GATT and NAFTA, states commit themselves to three fundamental rules: the most-favoured-nation treatment principle, the national treatment principle, and finally the prohibition of quantitative restrictions on import or export of products[2]. These rules were essentially designed to limit protectionism. Controversies over trade and environment usually concern state regulations conducive to import restrictions. The question is whether these environmental regulations are or are not disguised protectionism (e.g. Shrimp-turtle case and Tuna-dolphin case). In the case of water resources, the discussions concern regulations designed to restrict exportation.

NAFTA rules are more restrictive than those of GATT. First of all, under article XI of GATT, 'duties, taxes or other charges' are allowed[3]. In theory, it could be envisaged under GATT (but not under NAFTA) that a country, in order to protect its water resources, raises its export taxes in such a way to dissuade importing countries. It should, however, be recalled that the general objective of GATT in the long term is to decrease and eliminate such tariffs. There exist two types of exceptions to the prohibitions or restrictions of export: those of article XI.2.a and those of article XX.b. and g. The exceptions of article XI.2 are very limited and concern only: 'export prohibitions or restrictions temporarily applied to prevent or relieve *critical shortages* of foodstuffs or other products *essential* to the exporting contracting party' (article XI.2.a, emphasis added).

Among these provisions, it is the general exceptions of article XX.b. and g. that are the most discussed.

> Article XX: 'Subject to the requirement that such measures are not applied in a manner which would constitute a means of arbitrary or unjustifiable discrimination between countries where the same conditions prevail, or a disguised restriction on international trade, nothing in this Agreement shall be construed to prevent the adoption or enforcement by any contracting party of measures : [. . .]
>
> b. necessary to protect human, animal or plant life or health; [. . .]
>
> g. relating to the conservation of exhaustible natural resources if such measures are made effective in conjunction with restrictions on domestic production or consumption'.

---

[2] 'Article XI: General Elimination of Quantitative Restrictions. 1- No prohibitions or restrictions other than duties, taxes or other charges, whether made effective through quotas, import or export licences or other measures, shall be instituted or maintained by any contracting party on the importation of any product of the territory of any other contracting party or on the exportation or sale for export of any product destined for the territory of any other contracting party'. . . .

[3] Which is not the case under NAFTA : article 314 : prohibition of duty, tax or other charge on export of a good.

To some extent it is the trade community, and more precisely the powerful Appellate Body of the WTO, that has the power to decide of the legitimacy of national conservatory measures. Indeed, international trade law is the most enforced field of international law. The WTO dispute settlement holds means to render its decisions compulsory.

Legal studies are very important to understand what is at stake with the recognition of water as an international commodity, but international water transfers and exports raise not only the legal question of trade rules but also, and maybe more fundamentally, the question of the property regime governing water resources. Basically, to whom does water in its natural state belong? Who can legitimately make decisions regarding water resources management?

For us, the important question is not when - technically speaking - water becomes a 'product' but rather whether water has been appropriated or not, and which rules govern it. In other words, it is necessary to distinguish between trade in water and trade in water rights. Interestingly, Cossy (2005) suggests that further studies should focus on the different national statuses of water resources. Indeed, the property regime governing water resources is too often overlooked in the literature on water and international trade law.

As we saw in the first part, to define a property regime equals defining the web of interests around the resource. In the case of international watercourses, various levels of interest are concerned. The problem is that it is not always easy to clarify how to coordinate and link these various levels of interest. Property regimes of international watercourses are usually defined at the national level but not at the international level.

As we saw in the former chapter, the definition of the property regime at the international level is often missing or weakly implemented. Because this level of definition if often missing, international trade rules applies by default. This argument bears a resemblance to Chichilnisky's demonstration on north-south trade and the global environment (Chichilnisky 1994). According to her, environmental degradation concerns the South more than the North because the property regimes over natural resources are more poorly defined in the South than in the North. She explains, 'ill-defined property rights alone can create trade,' and that international trade under these conditions exacerbate environmental degradation and natural resources overuse in the South.

For her, however, a well-defined property regime is necessarily a private property rights regime. In any case, her demonstration has the merit of highlighting the fundamental aspects of property regime in the discussion over trade and the environment.

The Great Lakes have recently been the subject of a huge controversy because of various projects of water exports. The issue debated was whether the existing rules governing the Great Lakes were sufficient to face legal claims under

NAFTA. This question can be interpreted as a question over the relationship and hierarchy between international water law (in this case, the 1909 Boundary Treaty) and international trade law (NAFTA). But, more fundamentally, the example of the Great Lakes reveals the challenge of defining the property regime of an international watercourse involving many levels of interest. In order to respond to trade claims over the Great Lakes, governors, premiers, and federal decision makers had to negotiate over new rules and an institutional design to manage Great Lakes water withdrawals. In other words, they had to clarify the property regime of the Great Lakes and agree on an adequate level of management and a decision-making procedure. These negotiations revealed the tensions between the federal governments and the state/provincial governments.

WATER DIVERSIONS FROM THE GREAT LAKES, THE 1909 BOUNDARY TREATY AND INTERNATIONAL TRADE LAW

It is interesting to spend some time studying the recent developments over the Great Lakes Basin. After years of negotiations, the Great Lakes governors and premiers finally signed in December 2005, an agreement in which they commit themselves to adopt measures to implement a common agreed standard for water diversions and withdrawals. Discussions over the design of an adequate legal framework for water resources management in the basin have been intense for many years. The key issue was over the respective role of and interaction between international trade rules and international water law.

Plans to divert water from the Great Lakes are not new. Since the 1950s, plans to move water out of the Great Lakes have been developed and often rejected because of political, economic and environmental costs. Among the most controversial plans, the Grand Canal Project was proposed by private sector interests in 1959. This scheme proposed connecting James Bay in Canada to the Great Lakes and then from the Lakes to the western United States. More recently, in the early 1980s, the US Corps of Engineers reviewed diversion plans to recharge the Ogallala High Plains Aquifer with Great Lakes water. Another controversial plan consisted in diverting water from Lake Superior to Montana for a coal slurry pipeline.

These plans, however, never reached the level of controversy raised by the Nova Group proposal to export bulk water by supertanker to Indonesia in 1998. For the first time, a private firm, the Nova Group based in Sault Sainte Marie, Ontario, asked a permit to sell Great Lakes water, i.e. to sell water of an international watercourse to a non-riparian country.

The Great Lakes' case is particularly interesting as it raises the question of the adequacy of the principles of international water law to respond to international water trade challenges. Following worldwide trends, rules developed to manage the Great Lakes first dealt with navigation problems (1909 Boundary Treaty), then water quality and pollution (1978 Quality Agreement) and more and more today with water quantity and consumptive use rights.

This part of our work is devoted to the analysis of the debates over the adoption of a legal framework for water management in the Great Lakes Basin, especially following the controversial demand of the Nova Group to export bulk water to Indonesia.

We will first look at the legal framework governing the boundary waters of the Great Lakes before these recent developments, and focus particularly on its ability to control water exports from the Lakes. Was it necessary to develop new measures to control water exports? What were the different options available? These different options will be evaluated through the analysis of two influential reports to decision-makers in the Great Lakes Basin: the 2000 International Joint Commission Report to the government of Canada and the USA and the 1999 Lochhead Report to the Great Lakes Council of Governors.

Institutional Framework of the 1909 Boundary Waters Treaty

Different Rules for Boundary Waters and Tributaries of Boundary Waters

As early as 1909 the government of the United Kingdom and the government of the USA agreed on the Boundary Waters Treaty. Beyond regulation for navigation, the Treaty provides rules for the uses, obstructions or diversions of boundary waters. Article III requires a special agreement between the governments of Canada and the United States, or approval of the International Joint Commission (IJC) for uses of boundary waters that 'affect levels or flows on the other side of the border'. It is important to mention that the Treaty does not deal with all waters of the Great Lakes Basin in the same way (see map below). The Boundary Waters Treaty applies only to 'boundary waters,' that is, 'waters [. . .] along which the international boundary between the United States and the Dominion of Canada passes'. Tributaries of boundary waters and transboundary rivers are governed by article II which provides that each nation has 'exclusive jurisdiction and control over the use and diversion' of these waters. However, the other party continues to have the right to seek legal remedies for any resulting injury.

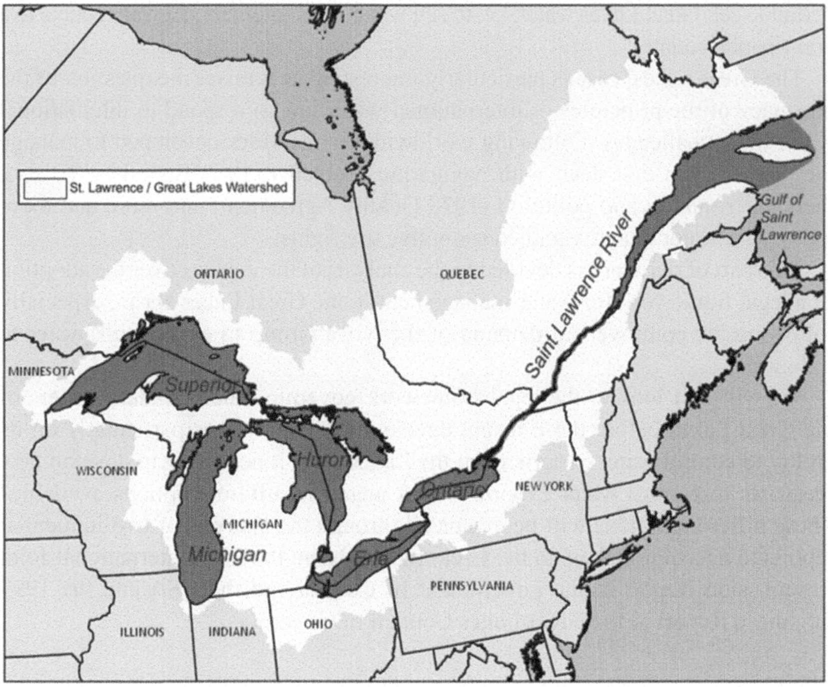

Figure 4 The Great Lakes- St. Lawrence River Basin (Credit to Karl Musser).

The Role of the IJC and its Guiding Principles

Article VII provides for the establishment of the International Joint Commission composed of six members, three from each country. Article VIII grants the commission 'jurisdiction over all cases involving the use or obstruction or diversion of boundary waters' where such uses 'affect the levels or flows of the waters' (article III). Article VIII defines also guidelines for decisions of the Commission. The parties have 'equal and similar rights in the use of boundary waters' and certain uses have priority over others: (1) uses for domestic and sanitary purposes; (2) uses for navigation, including the services of canals for the purposes of navigation; and (3) uses for power and for irrigation purposes. The IJC is granted four different powers: administrative, adjudicative (article VIII), investigative (article IX) and arbitral powers (article X) (Piper 1967, 81; Reinumägi 1985, 316). Under article IX one or the other party may request the

IJC 'to examine and report' on any questions or matters of difference between the parties. Article X grants arbitral power to the IJC. Under this article, and with the consent of the two parties, the Commission might be requested to 'decide' in case of a different between the two parties.

In sum, the main realization of the 1909 International Boundary Treaty is the creation of the IJC which was granted important powers. Over the last century, the IJC has been a significant tool for cooperation between the US and Canada and had a well respected role in the Great Lakes Basin and elsewhere along the border.

The IJC Reports on Great Lakes Diversions and Their Critiques

The IJC has been requested under article IX to report on the potential effects of water diversions on the Great Lakes at least twice. A first report on diversion was released in 1985, and a more recent one in 2000 responding to states' concerns over water export projects from the Great Lakes.

In the debates over the methods to prohibit water diversions from the Great Lakes Basin, the authority of the IJC itself has been subject to criticism on the US side. However, Canada's strategy to prohibit water exports from the Great Lakes consisted primarily in amending the International Boundary Waters Treaty Act, manifesting thus a clear will to continue to use the bi-national legal framework and international water law to face the emerging problems of water exports. The Canadian federal government decided to legislate on prohibition of 'bulk water removal from river basin' instead of prohibition of 'water exports'. The latter prohibition was supposed to go against Canada's international trade obligation whereas the first approach, by protecting water 'in its natural state', before it has become a commercial good, was considered to enhance environmental and conservation objectives and thus as being consistent with trade obligations.

Acting in the context of the 1909 Boundary Waters Treaty and using similar phrasing to the text of the 1909 Treaty, Canada's amendment prohibits bulk removal of boundary waters with few exceptions and provides for the establishment of a licensing system for projects of '*use, obstruction or diversion*' that are likely '*to affect the natural level or flow of boundary waters, waters flowing from boundary waters, or transboundary waters on the other side of the border*'. Besides introducing the bill in the House of Commons and declaring a temporary moratorium on water exports - while Canadian provinces adopt adequate legislation over Canadian waters - the Canadian government requested, along with the US, a report to the IJC on water exports from the Great Lakes. In sum, the strategy of the Canadian government clearly shows its intention to continue to work within the IJC jurisdiction.

It is interesting to note that IJC 1985 final, 1999 interim and 2000 final reports to the US and Canada have basically the same conclusions despite the evolution of the international context and notably the development of international trade rules. In the letter of reference ordering the 1985 report, the US and Canadian governments asked the Commission 'to examine into and report upon the effects of existing and proposed diversions within, into or out of the Great Lakes Basin, and the effects of consumptive uses on Great Lakes water levels and flows.' In the 1999 letter of reference, the governments recalled the results of the 1985 report but explicitly mentioned their alarm over the water export proposals and potential cumulative impact on the Lakes. Implicitly evoking trade obligations, they wrote:

> 'The governments are concerned that current management principles and conservation measures may be inadequate to ensure the future sustainable use of our shared waters' (1999 Letter of Reference).

In 1985 and 1999, the IJC has been basically advocating for the continuation to act within the framework of the 1909 Boundary Treaty and international water law principles. And indeed, as described earlier, the Boundary Treaty does provide some regulation for 'uses, obstruction and diversions of boundary waters.' The IJC even recalls that treaties are not the only source of international law but that 'agreements, custom, judicial decisions and writings' as well as 'the jurisprudence of the IJC' provide useful guidance to decide on diversions issues[4]. Concerning small and large diversion concerns, the IJC further mentions that the US and Canadian governments could well imagine working on a specific agreement over water quantity and use in the framework of the 1909 Boundary Waters Treaty as was done in 1978 over water quality problems. 'Article IV of the Boundary Waters

---

[4] 'The fact that the Boundary Waters Treaty does not deal specifically with water use questions of current concern does not mean that such matters have been or can be ignored, any more than it can be said that the Boundary Waters Treaty is insufficient for the contemporary setting. [. . .] Any consideration of matters pertaining to Great Lakes waters requires an awareness of the scope and content of the relevant international legal regime. In this regard, the significance and accomplishments of the Boundary Waters Treaty must not be under-estimated. The overall international legal regime is not, however, to be found only in the texts of treaties. It has evolved and continues to evolve through a combination of agreements, custom, judicial decisions and writings. The jurisprudence of the International Joint Commission is a particularly significant element. In addition, it is necessary to look at history in order to put the various elements in their proper perspective.' IJC 1985, 41.

Treaty contains a simple but powerful commitment by the two governments [. . .]. From this provision has emerged the more detailed Great Lakes Water Quality Agreements along with associated institutional mechanisms in both countries' (IJC 1985, 42). For the IJC, such a procedure could well be applicable to quantity, use or diversions matters on the Great Lakes.

Its recommendations consist essentially in establishing bi-national institutions to collect data, monitor, project and review potential impacts of existing and new diversions and consumptive uses in the Great Lakes. The IJC further recommends the review of current public polices at the federal and state/provincial levels to identify best practices.

Despite the emerging trend toward international water exports, in its 1999 interim report the IJC adopts the same reasoning as in 1985. It continues to advocate for a bi-national framework, the implementation of the 1909 Boundary Treaty and international water law principles. The Commission even suggests that it be given a central role in the monitoring of the implementation of its recommendations (IJC 1999, 30). Finally, encouragements to regional cooperation through the implementation of the 1985 Great Lakes Charter - a regional good faith agreement between Great Lakes governors and premiers - are made recalling the wider international framework. '*Without prejudice to the authority of the federal governments of the United States and Canada*, the governments of the Great Lakes states and Ontario and Quebec should not permit any proposal for removal of water from the Great Lakes . . . .' (IJC 1999, 29, emphasis added).

Reaction of the Great Lakes Governors[5] to the IJC Interim Report

It is interesting to analyze the reaction of the Great Lakes Governors to the IJC interim report. In a letter written a few days after the release of the interim report in August 1999, the governors communicated a set of principles under which the changes in the water management of the Great Lakes should be based. Mentioning the main conclusions of a report written by legal experts for the Council of Governors - the Lochhead report - the governors criticize indeed 'the effectiveness, adequacy and durability of the current regime' to face international trade legal challenges and therefore propose the design of a new legal standard able to endure trade challenges and based on the authority of the Great Lakes governors and

[5] The Council of Great Lakes Governors was formed in 1983 by the Governors of Illinois, Indiana, Michigan, Minnesota, Ohio, and Wisconsin to serve as a forum for the development of regional agreements between the Great Lakes Governors on issues of common concern : economic development, protection of the Great Lakes Basin, pollution problems, etc. In 1989, the Council Governors were joined by the Governors of New York and Pennsylvania.

premiers instead of that of the federal governments. The governors wrote: 'the legal standard [. . .] must vest authority in the Basin. Decision-making must be vested in those sovereign authorities to those of us, the Great Lakes governors and premiers, who manage the resource on a day-to-day basis.' According to the governors, 'the state and provincial governments are the best equipped to protect the Great Lakes.' It is worth mentioning that the letter was signed by the eight Great Lakes governors only without the premiers of Quebec and Ontario. It is not surprising if one recalls the Canadian strategy to put diversions and water management of the Great Lakes under the authority of the IJC.

For Canada, indeed, the framework of the 1909 Boundary Treaty with the IJC as authoritative body offers a decision-making procedure more equitable than that of a regional body. In a regional setting, the voting power of Canada would be in minority in comparison with the US power because the Great Lakes Basin is formed by eight US states but only two Canadian provinces, whereas the IJC is composed equally of three members from each country.

Beyond the voting issue, what is at stake in the discussion over the level of authority and management is the relationship with international trade obligations. To choose to regulate water diversions through an international agreement tightly linked to the 1909 Boundary Treaty - as was done for water quality in the 1978 Agreement – would correspond to develop a parallel legal regime independent from international trade law. In some ways, the relationship between such a Treaty and international trade obligations would be similar to the complex relationship between Multilateral Environmental Agreements (MEA) and international trade law.

In its 2000 final report, the IJC maintained most of its recommendations despite governors' criticism.

> 'The Boundary Waters Treaty, after 90 years, continues to provide effective protection for both countries from abuses to the waters of the Great Lakes Basin ecosystem. It represents a proven regime for avoiding and resolving disputes that arise between Canada and the United States over boundary waters and transboundary rivers' (IJC 2000).

The IJC seems to justify even further the needs to act at the federal and bi-national levels rather than at the regional level. The IJC presents the drawbacks of the Great Lakes Charter and domestic legal regimes to face potential trade challenges. The IJC notes that the Great Lakes Charter is non-binding, sets too high amounts to trigger other states consultation and does not require the consent of all Great Lakes states and provinces before allowing a diversion. Concerning the dormant commerce clause doctrine, the IJC mentions: 'the dormant commerce clause doctrine could be a constitutional restraint on state efforts, as opposed to *federal efforts*, to protect the resources of the Great Lakes' [. . .]. The potential restraint is reduced considerably if the states can agree on common standards for the use and protection of Great Lakes waters and can coordinate their water-management programs with *the federal and bi-national efforts*' (IJC 2000, emphasis added). Further 'there are now laws in both countries that, in different ways, limit removals of water from the Great Lakes Basin. These laws, however, apply only in the jurisdictions that enacted them; can be changed by those jurisdictions at any time and do not constitute a bi-national regime'. The IJC mentions further that current problems of diversions in the Great Lakes have important similarities with other issues in border waters outside the Great Lakes Basin. Regrouping research efforts on common concerns in the border region would avoid work to be duplicated and allow work to continue bi-nationally. Thus, the Commission reaffirmed the necessity to continue to act at the federal and bi-national level.

Concerning trade issues, the Commission presented the diverging legal opinions it received – including the Lochhead report - but essentially maintained its position. In its 1985 analysis of the legal regime governing the Great Lakes diversions, the IJC did not mention any potential constrains linked to trade law obligations. Despite the fact that the report was issued three years after the US Supreme Court *Sporhase v. Nebraska,* in which the Court declared that water was an article of commerce, the Commission did not analyze the consequence of such a qualification on the Great Lakes' legal regime. The Commission's report only mentions:

> 'Though some trends leading to major change may be discernible now, their nature and scope are to a large extent unpredictable, for the Great Lakes and elsewhere. For example, change in attitudes or in economic imperatives could make water a widely accepted article of commerce. The impetus for change may come from within the basin, or it may originate well outside the affected area' (IJC 1985, 49).

For the IJC, at that time, Great Lakes water was not an article of commerce and should continue to be regulated by the 1909 Boundary Treaty as well as international water law principles.

In 1999 and 2000, the IJC's legal analysis of trade obligations on the Great Lakes is based on a clear distinction between two types of water. Water in 'its natural state' and water 'entered into commerce'. For the Commission, the first type of water, i.e. water 'in its natural state' is regulated by international water law, whereas the second type, i.e. water 'entered into commerce' or 'captured' is subject to international trade law[6]. For the IJC, 'it is unlikely that water in its

[6] 'Unless water, in any form, has entered into commerce and has become a good or product, it is not covered by the provisions of any trade agreement, including NAFTA ; and that international rights and obligations respecting water in its natural state are constrained in separate treaties, such as the Boundary Waters Treaty, negotiated for that purpose' (2000 IJC report).

natural state (e.g., in a lake, river, or aquifer) is included within the scope of any of these trade agreements since it is not a product or good. This view is supported by the fact that the NAFTA parties have issued a statement to this effect'. However, 'when water is 'captured' and enters into commerce, it may attract obligations under the GATT, the FTA, and the NAFTA' (IJC 2000).

The Commission, however, then presented several diverging legal experts' opinions on the subject in order to justify its caution with respect to international trade law. Therefore, the IJC stood back from its argumentation and preferred instead to highlight the current legal debates and 'uncertainties' of the outcomes of the matter.

Acting with caution, the IJC recommends that measures to preserve the Great Lakes be taken in accordance with international trade rules:

> 'International trade law obligations [. . .] do not prevent Canada and the United States from taking measures to protect their water resources and preserve the integrity of the Great Lakes Basin ecosystem. Such measures are not prohibited *so long as there is no discrimination by decision makers against persons from other countries in their application, and so long as water management policies are clearly articulated and consistently implemented so that undue expectations are not created*' (IJC 2000, emphasis added).

Thus, the IJC conclusions reflect some ambiguities and hesitations concerning the exact nature of the interaction between international water law principles and international trade obligations. It fails to make the distinction between trade in water and trade in water rights. As we will show later, the IJC's distinction between water 'in its natural state' and water 'entered into commerce' is probably not the most useful argument to discuss the application of trade law to water resources. If one studies more carefully the decisions of the US Supreme Court on water as an article of commerce, one discovers a more important distinction that could be applied to international water resources: the distinction between regulated and unregulated water or, in other words, between water and water rights.

The Lochhead Report to the Governors

The already mentioned Lochhead report presented a different interpretation of trade obligations on Great Lakes states and provinces than the IJC did. The so-called Lochhead report was written by a team of legal experts composed of six American and Canadian lawyers commissioned by the Council of Great Lakes Governors (Lochhead *et al.* 1999).

According to the legal expert team that wrote the Lochhead report, 'the ability of any authority - state, provincial, federal or bi-national - to impose outright prohibitions on water exports is constrained by US and Canadian constitutional and international trade law. An outright ban on water exports would likely constitute an impermissible restraint on international trade' (Lochhead *et al.* 1999, 3). For them, even an amendment to the Boundary Waters Treaty between the USA and Canada prohibiting water removals would constitute an impermissible restraint on international trade unless the measures are taken in *an even-handed manner*. According to Lochhead and the other experts of the team, the distinction between various types of water, i.e. 'water in its natural states' and 'water entered into commerce' could be easily challenged. Moreover, the 1993 Statement between the USA, Mexico and Canada does not resolve the problem of water exports by other means than bottled water. They argue 'the Statement does not directly address the status of water in a pipeline or tanker hold that may be treated like a good the same as bottled water' (Lochhead *et al.* 1999, 19).

Even more controversial is the report's conclusion concerning the US dormant commerce clause and Water Resources Development Act (WRDA).[7] Based on an analysis of US jurisprudence of *Sporhase v. Nebraska*, and *City of El Paso v. Reynolds I and II*[8], the Lochhead report explains that the dormant commerce clause imposes extremely severe limitations on a state seeking to impose restrictions on interstate transfer of water. They provide a strict interpretation of *Sporhase* in which the US Supreme Court found that groundwater was an article of commerce, subject to the dormant clause. They mention: 'The Court struck down Nebraska's reciprocity requirements in its water export statute, finding that 'there is no evidence that this restriction is narrowly tailored to the conservation and preservation rationale' (Lochhead *et al.* 1999, 16). Further, they recall the District Court decision in *City El Paso I*. 'The Court found that the embargo was not and could not be defended as necessary to promote the health and safety of New Mexico's citizen because New Mexico faced no current or foreseeable shortage of water'. And finally, in *El Paso II*, the 'Court stated that any denial of a permit to an out-of-state user under the statute would be unconstitutional if the denial was made to limit exports in order to minimize in-state shortages for local economic benefit' (Ibid., 17). Without any clear argumentation, they declare: 'Great Lakes

[7] Water Resource Development Act 1986: 'No water shall be diverted from any portion of the Great Lakes within the United States, from any tributary within the United States of any of the Great Lakes, for use outside the Great Lakes Basin unless such diversion is approved by the Governor of each of the Great Lakes States'.

[8] These cases are discussed in the next section.

Basin water is even more likely than Nebraska and New Mexico groundwater to be held an article of commerce subject to the commerce clause.'

In sum, Lochhead's legal team provided a threatening and one-sided interpretation of US jurisprudence for out-of-state water transfers statutes (Dellapenna 2000).

Although it is usually acknowledged that Congress can authorize states to impose otherwise impermissible burdens on interstate commerce, the Lochhead's report argues that 'because WRDA does not clearly and unambiguously indicate a Congressional intent and policy to allow Great Lakes Basin governors to violate the commerce clause in vetoing a proposed out-of-basin water diversion, a veto under WRDA as it now stands would probably not survive challenge under the dormant commerce clause' (Lochhead *et al.* 1999).

Thus, the report concludes on 'the inadequacy of existing institutions and legal frameworks' (i.e. Boundary Treaty, Great Lakes Charter, WRDA) and advises that the new regulation of water withdrawals of the Great Lakes states be even-handed, be equally applicable to in- and out-of-basin uses and be tied to the protection and conservation of the Basin's water-dependent natural resources.

The analysis of the IJC's report and Lochhead's report shows the profound disagreements, and therefore uncertainties over the extent to which trade rules apply to Great Lakes water. As there has been to date no GATT or NAFTA or dormant commerce clause disputes to address restrictions on water exports from the Great Lakes, it is difficult to provide clear conclusions over the issue. What is certain is that the Lochhead report heavily influenced the Council of Great Lakes Governors, who decided to adopt a new standard in order to face potential trade challenges. The new legal strategy was released by the Great Lakes governors in the so-called Annex 2001, an annex to the 1985 Great Lakes Charter.

Toward the Marginalization of the IJC

Thus, contrary to what the IJC recommended, but following Lochhead's advice, the Great Lakes governors decided to develop a new institutional framework at the regional level instead of at the federal and bi-national level. As already mentioned, this choice led to marginalization of IJC authority to regulate Great Lakes diversions. Besides the different understandings of trade obligations on Great Lakes water, a reason for that choice is the controversial status of Lake Michigan which is not defined as boundary waters by the 1909 Boundary Treaty and on which the Great Lakes states want to keep full control. The Boundary Waters Treaty applies indeed only to 'boundary waters' that is 'waters [. . .] along which the international boundary between the United States and the Dominion of Canada passes'[9]. Although interconnected with the rest of the basin, Lake Michigan is considered to be US waters as it lies entirely inside the US territory.

This is precisely this issue that initially led to the preference for regional water management and thus the negotiation of the Great Lakes Charter in 1985. In 1986, the US Congress adopting the Water Resources Development Act (WRDA) recognized the central role of Great Lakes states and provinces to manage Great Lakes water. These two documents created the foundation for the development of a parallel institutional framework that recently - with the signature of the 2005 Agreement between the eight US Great Lakes governors and the two Canadian premiers - seriously marginalized the role of the IJC in the basin.

The different strategies adopted in reaction to the controversial Nova Group water export proposal are a good illustration of these two parallel institutional designs. As already mentioned, the Canadian federal government reacted by amending the Boundary Waters Treaty Act and by requesting the IJC for recommendations on water exports matters. In the US, arguing that Michigan waters were not included in the 1909 Boundary Waters Treaty, a Senator from Michigan introduced a bill to amend the 1986 WRDA in order to explicitly prohibit out-of-basin water '*exports*' (instead of 'diversions' only). The WRDA is a federal law which grants a veto power to Great Lakes states on water diversions and exports. It also encouraged the Great Lakes states, in consultation with Ontario and Quebec, to develop and implement a common conservation standard for making decisions concerning the withdrawal and use of water from the Great Lakes Basin in the spirit of the Great Lakes Charter.

In response to WRDA 2000, the Great Lakes governors and premiers of Ontario and Quebec signed the Great Lakes Charter Annex 2001. The Great Lakes Charter Annex is a non-binding agreement establishing a series of principles for a new common, resource-based conservation standard. The purpose of the Annex is to create a process for protecting the Great Lakes Basin from water withdrawals and diversions, which is consistent with the provisions of the US dormant commerce clause, and international trade obligations. Thus, clearly, Annex 2001 principles reflect the findings of Lochhead's report. Since 2001, the premiers and governors

---

[9] 1909 Boundary Waters Treaty preliminary article: 'For the purpose of this treaty boundary waters are defined as the waters from main shore to main shore of the lakes and rivers and connecting waterways, or the portions thereof, *along which the international boundary between the United States and the Dominion of Canada passes*, including all bays, arms, and inlets thereof, but not including tributary waters which in their natural channels would flow into such lakes, rivers, and waterways, or waters flowing from such lakes, rivers, and waterways, or the waters of rivers flowing across the boundary' (preliminary article 1909 BWT, emphasis added).

of the Great Lakes have been negotiating on an Annex Implementing Agreement, which was finally adopted in December 2005.

Negotiation of the 2005 Great Lakes-Saint Lawrence Agreement and Compact

Reflecting Lochhead's report, a major characteristic of Annex 2001 is the fact that the new standard will apply to all water withdrawals, not just diversions (i.e. even-handed measure to in and out-of-basin uses). A second principle derived from Lochhead's analysis is the 'benefit' or 'improvement' standard[10].

The first implementing agreement draft released for public review in July 2004 was rejected, notably by the province of Ontario. Overall, the final agreement is very different from the 2004 draft proposal. Different controversial principles such as improvement standard were finally suppressed in the final 2005 Agreement. The 'principle of equal and uniform treatment to all water users [. . .] without regard to established political boundaries' was also eliminated.

The final standard adopted on December 13, 2005 represents a compromise between the diverging views of trade obligations on Great Lakes water. It is a mix of the recommendation of the Lochhead's and IJC's reports. It is difficult to dissociate in the final agreement between both influences. Overall, the form of the agreement, i.e. state-province regional agreement and US compact, is a result of Lochhead's report and pressures from the Great Lakes governors. However, the core principle of the agreement, i.e. the protection of the integrity of the Great Lakes Basin ecosystem - used to justify the general prohibition of the diversions and to review exceptions - seems to be principally linked to the recommendation of the IJC.

The new framework consists of two separate agreements: the Great Lakes-Saint Lawrence River Basin Sustainable Water Resources Agreement between the Great Lakes states and provinces and the Great Lakes-Saint Lawrence River Basin Water Resources Compact among the eight Great Lakes states. The provinces-states international agreement is 'a good faith agreement': the Parties committed to render it binding by adopting appropriate domestic laws, which was done by 2008. In the US, the Compact was approved by Congress and signed by the President in October 2008.

---

[10] 'The standard that most effectively protects the Great Lakes and that best promotes the sustainable use of water in the future is one premised on benefit to the resource' (Lochhead *et al.* 1999, 45).

The Agreement and Compact declare a general prohibition of diversions (article 200) with few exceptions (article 201). But the Agreement and Compact do not address new or increased diversions alone but also new and increased in-basin withdrawals and consumptive uses. In this sense, they represent an attempt to create even-handed measures susceptible to face potential trade challenges. The decision-making standard is indeed very similar for uses in the same watershed and excepted diversions. Diversions are generally prohibited to the exception of diversions toward straddling communities (for public water supply purposes only), straddling counties (for public water supply purposes only) and intra-basin transfers. Among other things, the standard for withdrawals and consumptive uses (article 203) and the exception standard for diversions (article 201) use the same language requiring both that 'all water withdrawn shall be returned [. . .] to the Source Watershed less an allowance for consumptive uses', the proposal shall result in 'no significant individual or cumulative adverse impacts', the proposal 'shall be implemented so as to incorporate environmentally-sound and economically-feasible water conservation measures' and the proposal shall be 'reasonable', i.e. it provides 'for efficient use of the water and will avoid or minimize the waste of water' (article 201 and article 203).

However, contrary to uses in the same watershed, diversion proposals above a certain amount of water have to be reviewed by the Regional Body. The Regional Body is composed of the governor or premier of each of the Parties, or a person designated by each of them (article 400). Interestingly, the 'overarching principle' for reviewing proposals is 'the protection of the integrity of the Great Lakes-Saint Lawrence River Basin Ecosystem' (article 500).

It is important to contrast this principle from those proposed in the Lochhead report. The principle of the protection of the ecosystem's integrity grants decision makers room for manoeuvre in evaluating the different uses in- and out-of the basin. The principle does not require that measures be even-handed for in- and out-of basin uses; they must, however, be consistent with the objective of the protection of the ecosystem.

By contrast, Lochhead proposed strict 'even-handed' standards 'based on benefit to the resource'. Considering that Great Lakes water is an article of commerce, the only solution – according to him - for the Great Lakes states and provinces to adopt a standard compatible with trade obligations is to adopt a standard based on benefit. In- and out- of basin uses should be assessed after their potential to increase 'benefits' to the region. Thus, for him, uses offering more benefits, i.e. more efficient uses should be privileged over less efficient uses.

After reviewing the proposal according to the ecosystem integrity principles, the Regional Body issues a declaration of findings that is 'considered' by the originating party (article 506.9). In other words, the Regional Body's decisions are not binding on the parties.

With the creation of the Regional Body and regional review procedure, the authority of the IJC and the powers of the federal US and Canadian governments are weakened. Besides, the dispute settlement provided in the 2005 Agreement does not refer to the arbitral power of the IJC nor does it provide for rules in case of parallel litigation.

A careful reading of the preamble of the 2005 Agreement reveals a clear intention to vest the responsibility of the Great Lakes water management into the Great Lakes states and provinces: 'The waters of the basin are a shared public treasure and the states and provinces as stewards have a shared duty to protect, conserve and manage these renewable but finite waters'. The main reason why the 2005 Agreement vests the power of decision to the Great Lakes states and provinces – and not to the federal governments- is the will of the US riparian states to keep their sovereignty over Lake Michigan. It is not surprising therefore that there is no recognition in the preamble of the international status of the Great Lakes Basin[11]. Although Lake Michigan is linked to the Great Lakes Basin, it is not considered as international water because it is located wholly within the US border.

It is interesting to mention that the 1978 Great Lakes Water Quality Agreement between the US and Canadian federal governments concerned the whole Great Lakes Basin and therefore included Lake Michigan. In 1909, the IJC powers were limited to the 'boundary waters' but the 1978 Water Quality Agreement extended the power of the IJC to the whole Great Lakes Basin, including Lake Michigan.

Thus, the shift of power from the federal level to the regional level is even more significant that there already exists a precedent in the region to integrate the whole basin into a US-Canada international agreement. Instead, in 2005, the negotiations resulted in an agreement among the eight Great Lakes US states and the two Canadian provinces.

Conclusion

One observes a clear shift of power in the Great Lakes Basin. The role of the IJC along with that of the Canadian and US federal governments has been marginalized in the decision-making procedures over Great Lakes water uses and withdrawals.

---

[11] 'The Great Lakes is a unique natural resource shared between Canada and the United States. Although situated on an international boundary, the Great Lakes are not international waters. Rather, primary control over water withdrawals rests with the states and provinces, and any solution to the problem must center on the primary role in regulating the resource held by the states and provinces' (Lochhead 2000).

It is the Great Lakes states and provinces that have the primary responsibility in deciding for Great Lakes water uses. This arrangement is the result of the clarification of the property regime of the Great Lakes Basin. The basin states and provinces feared that their interests would be threatened if the decision-making power over the Great Lakes water belonged to the two federal governments.

The rationale used to justify the general prohibition of diversions from the Great Lakes Basin is the protection of the integrity of its ecosystem. The concept of ecosystem integrity has the double merit of being 'NAFTA-proof' and useful to balance federal against local public interests.

In the case of the Great Lakes, the fear concerned not so much international water exports but rather important diversions toward the South West of the United States. In this sense, the 2005 Agreement is the result of fear of attack under the commerce clause. Under US law, water is indeed 'an article of commerce'. What does it really mean? To what extent were the Great Lakes states regulations threatened by this qualification?

WATER AS AN ARTICLE OF COMMERCE IN US LAW

In 1982, the US Supreme Court in *Sporhase v. Nebraska ex rel. Douglas* qualified, for the first time, groundwater as an article of commerce and applied the US dormant commerce clause to Nebraska Water Export Statute. Since this decision of the US Supreme Court has often been misinterpreted, it is important to study the entire reasoning of the Court. The Court highlights the various forms of water rights and the different levels of resource ownership in different jurisdictions. Thus, although the reasoning of the court is sometimes opaque, there is an important distinction to make between water itself and water rights (Corker 1983).

Sporhase can be interpreted as a conflict between state and federal levels of public interest. *Sporhase* concerns interstate waters. Water allocation from the Ogallala Aquifer is neither governed by an interstate agreement nor by a federal regulation. In this context, the absence of a definition of property regime at the interstate level, the Supreme Court – in order to recall the federal interest in this interstate aquifer - applied the commerce clause. Unable to find alternative federal regulations, the court applied the *Pike v. Bruce Church* test to balance local and federal public interests.

The US Dormant Commerce Clause

The commerce clause of the US Constitution states that 'Congress shall have . . . power . . . to regulate commerce . . . . among the several states' (Article I §8 US constitution). The Supreme Court inferred from this clause the doctrine of the so-called 'negative', 'silent' or 'dormant' commerce clause. It limits state power to legislate in matter related to interstate commerce. Under the clause, a state may not discriminate against interstate commerce to advance the economic interests of the state or its citizens.

The Court applies basically two standards to evaluate state law, depending on if the state regulation is 'facially discriminatory' or not. If the regulation is 'facially discriminatory', the Court applies a 'strict scrutiny' test[12] under which only if state law was necessary and no other alternative was available, or if state law corresponds to an important 'legitimate local public interest', would the law be accepted. If the law is not 'facially discriminatory', the Court uses a balancing test to evaluate the purpose and effects of the measure[13]. The legitimate local public interest is balanced against the burden imposed by the law on interstate commerce. Two exceptions are usually recognized: if Congress approves the law and 'if a state is acting as a market participant, rather than as a market regulator'.

The Sporhase Case

The facts

In *Sporhase v. Nebraska ex rel. Douglas*, the Supreme Court qualified groundwater as an article of commerce and therefore extended the dormant commerce clause to water, raising great concerns. The controversy was huge and some feared that such a decision would threaten the equitable apportionment rules as the waters in question were part of an interstate underground aquifer shared by Colorado, Nebraska, Texas, New Mexico, Oklahoma and Kansas: the Ogallala Aquifer.

In 1972, Mr Sporhase, a Colorado resident, bought a farm whose land covered 140 acres in Colorado and 500 acres across the state line in Nebraska. The farm had a well on the Nebraska side, registered with the Nebraska Department of Water Resources by the former owner in 1971. Sporhase used groundwater pumped from the well located on the Nebraska side of its property to irrigate his corn and bean fields in Colorado. In 1976 the Nebraska Department of Water Resources warned Sporhase that proceedings would be commenced against him unless he stopped pumping water and using it in Colorado without a Nebraska transfer permit. In 1977, Sporhase applied to Colorado for a permit to drill a well on the Colorado side, but was turned down by the Colorado officials. Colorado denied the permit because they determined that the aquifer was already overused

[12] *Hughes v. Oklahoma*, 441 US 322, 337 (1979).

[13] *Pike v. Bruce Church Inc*, 397 US 137 (1970).

in the area. Sporhase thus continued to transfer water pumped in Nebraska for use in Colorado without a Nebraska transfer permit.

Indeed, although Sporhase's well was registered to the Nebraska Water Resources Department, Sporhase needed a special permit to transfer water from Nebraska into another state. Nebraska statute provides that

> 'any person, firm, city, village, municipal corporation or any other entity intending to withdraw groundwater from any well or pit located in the state of Nebraska and transport it for use in an adjoining state shall apply to the Department of Water Resources for a permit to do so. If the director of water resources finds that the withdrawal requested is reasonable, is not contrary to the conservation and use of groundwater, and is not otherwise detrimental to the public welfare, he shall grant the permit *if the state in which the water is to be used grants reciprocal rights to withdraw and transport ground water from that state for use in the State of Nebraska*' (Nebraska statute §46-613.01, emphasis added).

As Colorado statute prohibits the exportation of its groundwater, Sporhase could never have obtained a permit to transfer water from Nebraska into Colorado. The Nebraska statute's reciprocity requirement could never have been fulfilled.

The trial court rejected Sporhase's defense based on the demonstration that the Nebraska statute imposed an undue burden on interstate commerce. After the Nebraska Supreme Court affirmed the case, Sporhase appealed to the Supreme Court of the United States.

The US Supreme Court's Decision

The US Supreme Court addressed three questions under the commerce clause: whether groundwater is an article of commerce; whether Nebraska statute imposes an impermissible burden on commerce; and whether Congress has granted special permission.

1-Is Groundwater an Article of Commerce?

Referring to a former decision by the US Supreme Court *Hudson County Water co v. McCarter* (1908), the Supreme Court of Nebraska held that groundwater was not an article of commerce but a 'great public good': the state 'finds itself in possession of what all admit to be a *great public good*, and what it has it may keep and give no one a reason for its will'[14]. The Nebraska Supreme Court ruled

[14] *Hudson County Water Co v. McCarter*, 209 US 349 (1908). In that case, the US Supreme Court sustained a New Jersey statute prohibiting interstate transfer of any surface water located within the state.

that under Nebraska law, groundwater is not a 'market item freely transferable for value among private parties, and therefore not an article of commerce'. *Hughes v. Oklahoma*'s decision is not applicable to water resources. Water resources are distinguishable from other natural resources because it is essential for human survival.

Sporhase's argumentation rested on a three-judge district court judgment *City of Altus v. Carr* in which the court found that groundwater was an article of commerce under Texas law and therefore that the Texas export ban statute was an impermissible burden on interstate commerce. They argued that this decision definitively recast the public ownership doctrine already rejected in *Hughes v. Oklahoma*.

Despite an explicit recognition of the different nature of water rights in Texas and Nebraska, and the acknowledgement of the multi-state character of groundwater, Justice Stevens - writing for the majority- held that groundwater was an article of commerce.

Different Nature of Water Rights in Texas and Nebraska

In *City of Altus v. Carr*, the district court found that under Texas law groundwater was an article of commerce. In Texas, water located under the land of a proprietor belongs to the proprietor. It is a rule of absolute ownership. In Nebraska however, the owner of a piece of land does not 'own' groundwater beneath its property. The land owner must register with the Department of Water Resources to get an authorization to withdraw groundwater. The Supreme Court of Nebraska explained that 'the owner of land is entitled to appropriate subterranean waters found under his land, but he cannot extract and appropriate them in excess of a reasonable use upon the land which he owns, especially if such use is injurious to others who have substantial rights to the waters, and if the natural underground supply is insufficient for all owners' (quoted in *Sporhase v. Nebraska*).

Thus the US Supreme Court recognizes that 'the surface owner who withdraws Nebraska groundwater *enjoys a lesser ownership interest* in the water than the captor . . . of groundwater in Texas'. But the Supreme Court argues further these differences do not remove Nebraska groundwater from commerce clause scrutiny 'for appellee's argument is still based on the *legal fiction of state ownership*' (emphasis added).

The Court was thus unable to conclude to Sporhase's private ownership of water resources located below his property, contrary to those located below the Mocks's territory in Texas (*City of Altus*). This conclusion, however, did not lead the Court to say that water in Nebraska is a public nor a state property (the Court seems to make no distinction between public and state property, both expressions are used alternatively).

The Court's demonstration that state ownership is a 'legal fiction' rests on the fact that water is transferred from rural to urban areas in Nebraska[15]. Although the appellee explains that these kinds of transfers are operated at a cost which represents the distribution cost only and not the value of the water itself, the Court insists that this reasoning does not demonstrate the public ownership of water resources. Indeed, the Court affirms further that the price of transferred water in those cases is equivalent to any other state regulation of price. 'A state's power to regulate prices or rates has never been thought to depend on public ownership of the controlled commodity'.

In addition, the Court demonstrates the federal interest in groundwater resources located in Nebraska. Recalling that 80% of water supplies are used for agricultural purposes, and that agricultural markets are of primary importance in interstate commerce, the Court concludes to the federal interest in Nebraska water issues (demonstration used also to nuance Nebraska's argument that water is 'essential to human survival'). On top of this, the multi-state character of groundwater resources in Nebraska and neighbouring states are used to justify federal power in the case. 'Groundwater overdraft is a national problem'.

In sum, the Court argues that 'appellee's claim that Nebraska groundwater is not an article of commerce goes too far: it would not only exempt Nebraska groundwater regulation from burden-on-commerce analysis, it would also curtail the affirmative power of Congress to implement its own policies concerning such regulation' (*Sporhase v. Nebraska*).

2- Does Nebraska Statute Violate the Dormant Commerce Clause?

The Court found that Nebraska's statute purpose to conserve and preserve diminishing sources of groundwater is 'unquestionably legitimate and highly important' and considered the three first conditions[16] of the statute to be constitutional even though they are not strictly even handled (This statute applies to interstate water transfers only and not to intrastate transfers). The Court accepts that there are 'legitimate reasons for the special treatment accorded requests to transport groundwater across state lines' (*Sporhase v. Nebraska*).

[15] 'The fiction is illustrated by municipal water supply arrangements pursuant to which groundwater is withdrawn from rural areas and transferred to urban areas' (*Sporhase v. Nebraska*).

[16] 'that the withdrawal of the groundwater requested is reasonable, is not contrary to the conservation and use of groundwater, and is not otherwise detrimental to the public welfare.' Nebraska Statute 46-613.01(1978).

The Court even acknowledges the legitimate role of the state in regulating water resources within its boundaries and agrees to a 'limited preference' for state citizens. 'Although appellee's claim to public ownership of Nebraska groundwater cannot justify a total denial of federal regulatory power, it may support a limited preference for its own citizens in the utilization of the resource .... In this regard, it is relevant that appellee's claim is logically more substantial than claims to public ownership of other natural resources' (*Sporhase v. Nebraska*).

Contrary to some radical interpretations of the case such as that of Lochhead, the Court clearly provides exception to the application of the dormant commerce clause in the case of water resources. Further, the Court mentions explicitly that water apportioned among states either by equitable apportionment decrees or an interstate compact is exempted from dormant commerce clause scrutiny. Indeed, 'the legal expectation that under certain circumstances each state may restrict water within its borders has been fostered over the years not only by our equitable apportionment decrees . . . but also by the negotiation and enforcement of interstate compacts. *Our law therefore has recognized the relevance of state boundaries in the allocation of scarce water resources'* (*Sporhase v. Nebraska*, emphasis added).

However, the Court found that the reciprocity requirement of Nebraska statute was facially discriminatory and could not be justified by conservation purposes. Finally, the Court was unable to find an explicit statement by Congress which would exempt Nebraska state from dormant commerce clause scrutiny. Thus, the Court concluded that the reciprocity requirement of Nebraska statute violates the commerce clause.

Balancing local and federal interests

Because *Sporhase* concerns an interstate aquifer, the Ogallala Aquifer, some authors feared that the commerce clause would threaten the equitable apportionment doctrine. The application of the dormant commerce clause to water resources corresponds for them to the replacement of equitable apportionment rules by free market principles erasing state lines (Williams 1983, Tarlock 1985). Are the dormant commerce clause and equitable apportionment doctrine really conflicting?

Tarlock argues in an important 1984 article that the equitable apportionment doctrine of the US Supreme Court is undergoing significant changes following cases such as *Sporhase* and *Vermejo*. He writes:

> 'States have long resisted subjecting water allocation to the harsh discipline of the market, but they are being forced to make hard choices about water. The era

> of federal reclamation projects and subsidies has ended, population growth will dictate substantial reallocations, and Supreme Court decisions such as *Sporhase* will spur the removal of market barriers as beneficial uses evolves toward the concept of economic efficiency. Thus, once the Court makes an equitable apportionment, further reallocations may be possible to an extent that would have been unthinkable in the past' (Tarlock 1985, 411).

Such interpretations go too far. We saw that in *Sporhase,* although the Court declared that groundwater was an article of commerce, it further recalled the particularity of that good and accepted exceptions to the application of the dormant commerce clause in the case of water resources (particularly water apportioned by interstate compacts or Supreme Court rulings). For the Court, groundwater is neither a pure private good nor a pure public good. The Court rejected the concept of 'state ownership' as a legal fiction: water 'ownership' means nothing other than who is in charge of making decisions about the resource. In the case of interstate groundwater, the Court found that the public nature of water resources involved different levels of responsibility, including individual, state and federal levels. Interstate water regulation is neither a pure federal power nor a pure state power. Using the *Pike v. Bruce Church* test to evaluate the regulation of water resources, the Court balanced local public interest with federal public interest. Unfortunately the Court did not discuss the distinction between public ownership and state ownership. One could argue that the Court rejected 'public ownership' only in the sense of 'state ownership' but not if the 'public ownership' included other types of interests than those of the state, i.e. federal interests as well.

In sum, the discussion over water as an article of commerce implies questions not of 'ownership,' but instead of who is in charge of making decisions about the resource and the type of public good it is: local or federal. In fact, this question of balancing legitimate local and federal interests is the same as in the case of the Great Lakes. To whom does Great Lakes water belong? That is to say, who is in charge of making decisions about Great Lakes water management and for which public interest, local or federal?

The answer to these questions regarding the Great Lakes is now given by the 2005 Agreement and Compact. The standard adopted for the management of water resources in the Great Lakes rests on the protection of the integrity of its ecosystem, and thus includes by itself clear limits to the scope of public interest. In the case of the Great Lakes, if both in- and out-of-basin uses are regulated, they do not follow the same standards. Indeed, the integrity of the ecosystem is a good benchmark to weigh the different types of interests at stake.

Whether water is an article of commerce depends on the water rights linked to it. In this sense, it is essential to have a good understanding of the true nature of the different individuals' water rights and the level of ownership of the resource corresponding to the right. We already mentioned the differences between Texas' water right in the *City of Altus* case and Nebraska's water right in *Sporhase*.

At the international level, it is important to take into account domestic status of water resources in order to evaluate the possibility of the recognition of water as a 'product' subject to international trade rules. However, as we will see, challenges to the current rules of international watercourses might come from investments rules such as those of chapter 11 of NAFTA rather than from the rules concerning trade in 'goods'. The first 'water' case in NAFTA was submitted under chapter 11. Sun Belt Inc filed a claim against Canada in 1998 with the UNCITRAL. The US firm challenged British Columbia water protection legislation and a moratorium on exports of bulk water from the Province. Among other things, under article 1110, Sun Belt alleged that British Columbia's regulations were tantamount to nationalization or expropriation of its investments. According to the Company, the case is still pending, while for the Canadian government it is inactive.

A more recent case involved Texan irrigators and the State of Mexico. Resting on a broad definition of investment, Texan irrigators argued that their water rights corresponded to investment in the meaning of NAFTA chapter 11. They argued further these 'investments' had been neglected by Mexican measures equivalent to an expropriation or nationalization (article 1110). In this case, the bilateral regime between Mexico and the US over the Rio Grande/Bravo was challenged by NAFTA trade and investment rules. The issue here is not bluntly whether water is a 'good' but whether Texan irrigators' water rights qualify for 'investment' under chapter 11 of NAFTA. The case raised the question whether these 'investments' should be considered as being located in Mexico or not.

INVESTMENT CLAUSE IN NAFTA, WATER RIGHTS AND INTERNATIONAL WATER LAW: THE BAYVIEW IRRIGATION DISTRICT V. MEXICO CASE

Historical Background

The lower part of the Rio Grande forms the border between the USA and Mexico - its main tributaries being located in Mexico (on the lower part). The 1944 Treaty between the USA and Mexico provides rules for the utilization of the Colorado and Tijuana Rivers and the Rio Grande/Bravo River. More specifically, the two states agreed on the allocation of water of the Rio Grande from Fort Quitman, Texas to the Gulf of Mexico (article 4). They also agreed to the construction of two international reservoirs – the Falcon and Amistad Reservoirs - on the Rio Grande to improve water management in the region (article 5). The provision that is relevant to the case is the allocation provided to the US on the Lower Rio Grande Basin. Article 4 (c) mentions that the US be allocated 'One third of the flow reaching the main channel of the river originating from the River Conchos, San Diego, San Rodrigo, Escondido, Salado and Arroyo Las Vacas [Mexican tributaries] and provides that this third shall not be less, as an average amount in cycles of five consecutive years, than 431,721,000 m^3 annually'. The article provides further specific rules 'in the event of extraordinary drought [. . .] making it difficult for Mexico to make available the run-off of 431,721,000 m^3 annually'. Finally, the Treaty expanded the functions and powers of the International Boundary Commission created in 1889 which became the International Boundary and Water Commission. The Commission is in charge of monitoring the implementation of the Treaty, preventing violation of the Treaty and settles disputes between the USA and Mexico.

At the end of the 1992-1997 cycle, following severe drought conditions, Mexico had a debt of 1,262 million m^3 (1.024 million acre feet). At the end of the following cycle, Mexico cumulated another important debt to the US. In February 2005, the two governments concluded an agreement over the Mexican 'debt' resulting from the 1992-1997 and 1997-2002 cycles. The cycles were declared closed.

Summary of the Dispute

Considering that their water entitlements corresponded to 'investments', Texan irrigators filed a claim under NAFTA chapter 11 against Mexico in August 2004. Chapter 11 applies to 'measures adopted or maintained by a Party relating to investors of another Party'.

Claimants' accusations

The claimants - seventeen irrigation districts of the Rio Grande Valley - denounced the holding and diversion by the Mexican government of a certain amount of water of the Rio Grande for the use of Mexican farmers. They considered that such an act was equivalent to an expropriation or nationalization of their investment by the Mexican government (article 1110 of NAFTA). They also claimed that Mexico treated the investment of US investors less favourably than it treated its own investors (article 1102 of NAFTA). The claimants argued that Mexico breached article 1105 under which each Party must grant the investments of investors from another Party treatment in accordance with international law, including fair and equitable treatment. According to the claimants, by intentionally violating the 1944 Treaty, Mexico clearly violated international law.

Irrigation Districts are US investors . . . .

The request for arbitration mentions: 'each claimant is an Investor and owner of an integrated Investment. This Investment includes, but is not limited to, rights to water located in Mexico; facilities to store and distribute this water for irrigation and domestic consumption; irrigated fields and farms; farm buildings and machinery; and ongoing irrigated agricultural businesses. [. .] Each claimant's investment is entirely predicated on this right to receive water located in Mexican tributaries. Such rights have long been recognized as property rights' (Marzulla 2005a, 27). The request mentions also that any 'Claimants' integrated investment meets the definition of investment contained in NAFTA. Article 1139(g) of NAFTA defines an investment as 'real estate or other property, tangible or intangible, acquired in the expectation or used for the purpose of economic benefit or other business purposes' (Ibid., 28).

. . .of investment located in Mexico

Although the claimants are investors located in Texas, USA, the case mentions that a fundamental part of their investments - that means their water rights - are located in Mexico. Indeed, according to the request for arbitration, 'Claimants, the *rightful owners* of the majority of the water rights allocated to the United States under the 1944 Treaty, were deprived of this water by measures taken by Mexico in violation of chapter 11 of NAFTA.' And further, 'Claimants are the *legal owners* of 1,219,521 acre-feet of the irrigation water wrongfully withheld and diverted from the Rio Grande by Mexico's manipulation of its dams and reservoirs as of October 2002 (Ibid., 36)'. Finally, the Claimants argue that Mexico violated chapter 11 when 'it captured, seized and diverted to the use of Mexican farmers the water located in Mexico and *owned by Claimants*' (Ibid., 38, emphasis added).

In a letter dated April 7, 2005 to the International Centre for the Settlement of Investment Disputes (ICSID), the claimants' attorney Marzulla comments on Mexican observations concerning the claim. The attorney justifies further its argumentation that Texas irrigation districts' water rights are investments located in Mexico. Referring to the Dublin Statement, the United Nations Conference on Environment and Development, various General Assembly resolutions, and recommendation (2001) 14 of the Committee of Ministers to Member States of the European Charter on Water Resources as well as the 'increasing support for the recognition of a right to water in numerous international instruments,' the attorney maintains that 'the international recognition of water as an economic good, to which individual entities may claim a right, confirms that claimants' request for arbitration should be registered for review by an arbitration panel' (Marzulla 2005 b, 4).

The Mexican defence

For Mexico, the claim fell outside the scope of NAFTA. Indeed, for Mexico, none of the claimants had an investment in Mexico. The water rights were not granted to the claimants by Mexican law but by Texas law. Mexico wondered 'how it is that Mexico could have expropriated a property right created by a foreign legal system and existing only in another system' (Mexico 2006, 34). Further, Mexico argued that 'the breaches of NAFTA alleged by the claimants are based exclusively on the argument that Mexico breached obligations established in the 1944 Treaty between the USA and Mexico'. [. . .] 'This tribunal does not have competence to find whether or not Mexico breached its obligation under the bilateral treaty and cannot presume such a situation' (Mexico 2006, 2). Besides, Mexico recalled the dispute settlement mechanisms provided in the 1944 Treaty to solve disputes over Rio Grande/Bravo. It also mentioned the recent negotiations over the Rio Bravo River 'debt' between the Mexican government and the US government and how they reached an agreement in February 2005. Thus for Mexico, the claim was untimely. In any case, the 1944 Treaty allocates water to states, and not to individuals.

The most important part of Mexico's memorial - for our concerns - is the one dealing with the nature of the rights to use water for irrigation granted to the Claimants by the US.

The Nature of Texas Irrigation Rights in the Rio Grande Valley

Under Texas law, a claimant's water rights – i.e. water rights administered by the Rio Grande Watermaster – correspond to a proportional share of the water available in the region and not to a specific quantity of water (see Part I, chapter 3).

Further, recalling the history of water legislation in Texas, Mexico refers to the important 1967 decision '*The States of Texas* et al., *v. Hidalgo County Water Control and Improvement District n°18* et al' in which the Texas Court of Appeal ruled that :

> 'the existence of the Treaty cannot be ignored as an important factual circumstance in this case, but the US did not by the Treaty confer on anyone a right emanating from the central government to make use of the waters of the Rio Grande.'

Thus, the Treaty of 1944 does not allocate any water rights to any individuals but to states only. It is then the states who are in charge of adopting rules to allocate the available water within their territory. The *Hidalgo* case makes clear, however, that : 'The State of Texas had neither control over nor official part in the construction of the Rio Grande storage dams nor does it regulate or control the release of waters from the Treaty reservoirs' (Mexico 2006, 19).

In fact, the riparian rights system and the prior appropriation system co-existed in Texas for many years. But, in the early 1950s, severe drought conditions led to many disputes over water rights. Besides, the completion of the Falcon Reservoir provided additional water to be apportioned. Thus, in 1956, the State of Texas initiated a court proceeding through which, in effect, it delegated its authority to allocate water use rights to the Texas Court. In 1967, the Texas legislature passed the Adjudication Act, which merged riparian and appropriation systems together on a state-wide basis and consolidated the water permit system. Water rights were to be administered by a watermaster. The official website of the section of surface water use of the Texas Commission on Environment Quality (TCEQ) mentions that surface water is state water and surface water may be used in most cases only with explicit permission of the state, except for domestic and livestock uses. It is further explained 'None of the water right documents guarantee that water will always be available, but some of them provide more certainty than others' (TECQ website).

In the Rio Grande Valley, water rights are monitored by the Rio Grande watermaster and are subject to specific rules established to regulate the Rio Grande Water Market (see Part I, chapter 3). Thus, in contrast with the rest of Texas, the priority of use does not depend on the emission date of the right but on the type of uses. In the Amistad and Falcon Reservoirs, municipal and industrial rights have priority over irrigation rights. The watermaster determines the available water for all uses after reviewing the information on storage levels at Amistad and Falcon Reservoirs, average inflows and outflows and the US share of these three variables provided by the International Boundary and Water Commission (Commission in charge of monitoring the 1944 Treaty).

Whereas water allocation for municipalities is decided at the beginning of each year, the water available for irrigation is reviewed at the beginning of each month.

In a 2004 study on water marketing in Texas by Environmental Defense, the authors record:

> 'Agriculture accounts for 85% of the water rights, but municipal, industrial and domestic use have a higher priority during dry periods. New inflows to the reservoir system [Falcon and Amistad Reservoirs] are first apportioned to municipalities based on an accounting system, otherwise known as the 'municipal reserve'. At present, the municipal reserve is set at 225,000 acre-feet, and it is replenished before any remaining inflows are allocated to agriculture. Thus, allocations to irrigators and irrigation districts can vary significantly according to climatic conditions. Between

> 1997 and 2002, when inflows into the Amistad/Falcon system from Mexico were greatly reduced, agricultural users were allocated, on average, 38% of their paper water rights' (Kelly 2004, 35).

Considering the rules governing irrigation water rights in the Rio Grande Valley, it seems difficult to support the argument that Texan irrigators are 'the rightful *owners* of the majority of the water rights allocated to the US under the 1944 Treaty' (Marzulla 2005 a, 32) nor 'the legal *owners* of 1,219,521 acre-feet of the irrigation water wrongfully withheld and diverted from the Rio Grande by Mexico' (Ibid., 36, emphasis added). Texan irrigators in the Rio Grande Valley do not own any part of the international waters of the Rio Grande but rather the USA, and, by delegation, the State of Texas.

Tribunal's Decision

The request for arbitration was filed on January 19, 2005 to the International Centre for the Settlement of Investment Disputes (ICSID). In July 2005, the ICSID certified the claim and announced that the arbitration could proceed. In December 2005, the Tribunal was constituted and the parties held the first session with the Tribunal in February 2006. In June 2007, the Tribunal released its decision. It found that it has no jurisdiction on the case because the claimants cannot be qualified as 'investors' under chapter 11 of NAFTA.

In paragraph 104 of its award, the Tribunal mentions:

> 'In this case the Tribunal does not consider that the claimants were 'foreign investors' in Mexico. Rather, they were domestic investors in Texas. The economic dependence of an enterprise upon supplies of good – in this case, water – from another state is not sufficient to make the dependent enterprise an 'investor' in that other state'.

Responding to the claimants' argument that the 1944 Treaty created property rights to US citizens, the Tribunal explains:

> 'The Tribunal can find no evidence in the 1944 Treaty to suggest that this imaginative interpretation of the Treaty, whose legal coherence and practical operability are open to considerable doubt, was intended by the Parties. The ordinary reading of the Treaty is that it is an agreement to apportion such waters as arrive in the international watercourse – the Rio Bravo/Rio Grande- between Mexico and the United States; and the Tribunal sees no reason whatever to doubt the correctness of that reading. [. . .] But the 1944 Treaty does not create property rights amounting to investments within the meaning of the NAFTA which US nationals themselves may protect by action under NAFTA Chapter Eleven'(ICSID 2007, §120-121).

Although the claimants later added to their main argument that water should be considered as a tradable good, the question here concerns investment provisions and not trade in goods. This case cannot be directly linked to the former discussions on water export and water as an article of commerce. It brings nevertheless interesting insight into our discussion of the various levels of interests and the property regime of international watercourses. The Bayview case reveals that the rules governing water rights on the Rio Grande in Texas law are valid within Texas only and not in the other jurisdictions sharing the same watercourse. The Bayview case highlights the existence of different levels of decision-making and management rules in international river basin. Thus, the property regime of an international watercourse has different 'layers'. Rio Grande water rights holders under Texas water law cannot claim that their rights are property rights directly resulting from the 1944 Treaty. The institutional framework and rules defined by the 1944 Treaty cannot be short-circuited. The Treaty makes allocations only to the two governments and not to their citizens.

In fact, the Bayview case is a good example of a two-level game (Putnam 1988). Texas water users tried to push the federal government to take a harder stance vis-à-vis Mexico, in vain, and therefore attempted to find an alternative way to put pressure on Mexico. In the late 1990s, in order to force the government of Mexico to comply with the 1944 Treaty, two Texan representatives submitted a Senate Congressional Resolution and a House Congressional Resolution. The Texas Commission on Environmental Quality (TCEQ) required that the US government terminate the 1944 Treaty or reduce their deliveries on the Colorado. These lobbying activities remained fruitless. Therefore, the Texan irrigators turned to NAFTA procedures. It is important to recall that issues on the Rio Grande are linked to two other watercourses shared with Mexico: the Colorado and the Tijuana Rivers. From the beginning of the relations between Mexico and the US on the Rio Grande, the provisions concerning the Rio Grande were linked to the allocation of the Colorado and Tijuana Rivers in the 1944 Treaty. It was thus difficult for the US government to renegotiate the provisions concerning the Rio Grande without reopening the discussion on the Colorado and Tijuana Rivers. Here again, one observes the different levels of interests in the international watercourse at stake. The public federal interest in the Rio Grande is conflicting with the local Texan interests. Who should decide water management in the Rio Grande Basin? In which public interests: bi-national, federal, regional, or local? Established by the 1944 Treaty, the International Boundary and Water Commission is a starting place for answering these questions. The power of the IBWC is however very limited. The two governments keep full sovereignty over the waters within their borders. For example, in its article 2, the 1944 Treaty provides:

> 'The works constructed, acquired or used in fulfilment of the provisions of this Treaty and located wholly within the territorial limits of either country although these works may be international in character, shall remain, except as herein otherwise specifically provided, under the exclusive jurisdiction and control of the Section of the Commission in whose country the works may be situated'.

The Treaty provides neither strong common decision-making procedures nor clear guiding principles for the common management of the shared watercourses. The Treaty instead allocates water quantity to the two sovereign states with limited flexibility only (five-year cycles).

More fundamentally, the Bayview case reveals the weaknesses of the 1944 Treaty to face current challenges resulting from differentiated evolution in water use, climate change and demographic growth. The 1944 Treaty does not provide an institutional framework able to resolve management problems in a common and adaptive way. There are no guidelines to coordinate and balance the different levels of public interests in the international watercourse.

The Bayview case is another example where trade rules have been used in order to advance particular interests in an international watercourse governed by poor common decision-making procedures and rules unable to adapt to changing conditions.

CONCLUSION

International water law is a fascinating field of international law. It is difficult however to generalize on international water law as its principles and rules vary widely with the geographical, economic and security conditions of each particular international river basin. Nevertheless it can be said that the ways current principles of international water law are enforced in many regions are insufficient to face the challenges linked to climate change, increased pressures on water resources, and uneven economic and demographic growth. As we observed at the domestic level, riparianism has evolved towards more formalized systems of water rights in many countries. Such an evolution at the international level requires first the enforcement of the common property regime of international watercourses. As we saw this means the establishment of well functioning common decision-making procedures and the clarification of management rules and principles. The definition of the common property regime of an international watercourse is not completed as long as the 'international layer' is not structured and the rules coordinating the different levels of public interest are not established. In the absence of such a structure, international trade law will continue to interfere with international water law. Trade rules are filling the legal gap resulting from the lack or weakness of the property regime governing international water resources.

The literature on water and international trade law usually overlooks the question of the property regime or legal status of water resources both at the domestic and international levels. In the discussions on water and trade law it is essential first to distinguish between appropriated and unappropriated water (or in other words, trade in water and trade in water rights). Then, it is important to take into account the different levels of ownership included in the different water rights. In fact very few jurisdictions in the world consider water as pure private good, being thus subject to trade rules. A rare example is groundwater in Texas. In most places, water rights are usually tied to a particular use, place and are limited in time. When water rights are transferable, there are usually regulations governing such transfers. Given these regulations, it is highly improbable that water will be considered a freely tradable good subject to international trade rules. Investment rules are another matter but they have less impact on decisions concerning international water transfers.

States do not need to ban water exports to keep control over their national water resources. Also, there is no reason to reject all international water transfers; there is only a need to define coherent criteria and guidelines to review transfer proposals. If such rules are lacking or are inadequate, trade rules will fill the gap.

For international watercourses, the challenge is harder. The adoption of common decision-making procedures and adaptive management rules is clearly complicated by the international nature of the watercourse. As we saw, the 'international layer' is often missing in the definition of the property regime of international watercourses, which results in increasing interferences of international trade law with international water law.

Part III

Market-Based Solutions along International Watercourses

Among the different policy options recognizing water as an economic good, water markets are increasingly presented as models for water reform worldwide. This enthusiasm for water market does not concern national water resources alone but also international watercourses. Thus, different proposals have been made for the creation of water markets along international watercourses as a way to contribute to international water disputes' resolution. However, whereas an important literature has emerged to discuss the worth of institutional designs for water market at the domestic level, there is no such literature for water markets at the international level. In fact there is no clear understanding of what an international water market would look like, what it would really imply and what it would require. To what extent and under which conditions could market-based mechanisms improve international watercourses management? Although market-based instruments could be used to manage pollution and other qualitative problems along international watercourses, our analysis focuses essentially on quantitative aspects.

As we saw in the first part, water markets are usually local programs and are, most of the time, located along a watercourse. Water markets indeed require

ISBN: 9781843393610. Published by IWA Publishing, London, UK.

specific conditions to emerge, including, at the least, conveyance infrastructure, and, because it is very expensive to move water from one place to another, rivers are usually used as natural conveyance facilities. International water markets are thus here understood as an institutional design monitoring the international trade in water rights along an international watercourse. International water markets are not understood here as including long distance bulk water exports by supertanker, water bags or pipelines. As we saw in the former part, international bulk water exports and transfers are likely to remain limited tools for water management worldwide, used on a case by case basis alone. Although their future is uncertain, they bring interesting questions and reveal the lack of definition of the property regime of international watercourses. Here, international water markets do not concern trade in water but trade in water rights.

International water markets have also been studied from the point of view of international trade in virtual water. This is not our objective here. It is true that the literature on 'virtual water' offers an alternative perspective on international water disputes. Virtual water being defined as the quantity of water used in the production processes of various industrial and agricultural products (Allan 1997, Hoekstra 2003). For virtual water advocates, because water is essentially used to produce crops, water problems can be eased by importing food products instead of producing them. Although this notion is a good way to put the water problems into perspective, we believe they do not resolve the fundamental problématique of water management.

In this third part, after reviewing international market proposals, especially that of Fisher (2005) - the most serious and recent one - we will turn to two case studies. Although not international – as no international water market for the time being exists - the Colorado River Basin and the Murray-Darling Basin offer very interesting elements to better understand the challenges raised by the introduction of interstate water trade. Indeed, in both cases, the river crosses different jurisdictional boundaries and, in a sense, federal states negotiations over interstate rivers have something to tell us about true international negotiations over international watercourses. Of course, much prudence in the transposition of the issues from the federal level to the international level is required.

Chapter 6

Economic Approaches to International Water Disputes

Economic approaches have been increasingly applied to water disputes among countries. The role of water economics in conflict resolution is, however, a controversial matter. As we will see, the recognition of water as an economic good is more challenging in the international context than in domestic politics. While for some economists the economic valuation of water resources can contribute to the resolution of water disputes, their studies however often neglect - more or less explicitly - legal and institutional aspects of water management across borders (E. Kally 1993, Fishelson 1994, Dinar and Wolf 1994, Zeitouni, Becker and Shechter 1994). How can we determine the economic value of international water resources? To what extent can water markets contribute to a better allocation of water resources among countries? Who should be entitled to participate in such a market – individuals or states? What kind of institutional framework would that require?

No matter what type of economic approach is adopted, the fundamental question of the initial allocation of water rights among countries has to be addressed.

In this regard, it is interesting to analyse the work of Franklin M. Fisher and his co-authors, who propose a 'new economic approach' for 'conflict resolution in the

ISBN: 9781843393610. Published by IWA Publishing, London, UK.

Middle East and beyond' (Fisher *et al.* 2005). By monetizing the value of water resources, they intend to deemphasize the question of property rights' allocation among conflicting parties and thus facilitate negotiations.

The main foundation of Fisher *et al.*'s 'new approach' is to 'deemotionalize the dispute' by monetizing the value of water resources in the region. Building on Fishelson's analysis, they explain that water scarcity is not a problem of quantity but a problem of cost. Indeed, according to Fishelson, the cost of desalination (plus conveyance costs) should be considered as an upper bound on the value of water in the region (Fishelson1994). Fisher *et al.* note, 'it is irrational to value water at more than the cost of replacing it. Hence the possibility of seawater desalination places an upper bound on the value of water (Fisher *et al.* 2005, 4).' 'It means, for example, that the value of water in dispute between Israelis and Palestinians is considerably less than $100 million per year (Ibid., 62)'[1].

The cost of desalination is an upper bound value but is not the current value of water in the region. In fact, the value of water for the coming years is calculated through the use of their Water Allocation System (WAS) model. The WAS is a tool designed to help decision makers adopt water policies and study different scenarios of water negotiations. It has thus both fields of application, the domestic and international arena. Fisher *et al.* summarize the description of WAS in the following manner:

> 'WAS divides the area to be studied into districts. Within each district, data are provided on water sources and costs of extraction, and demand curves are specified for three user classes: households, industry, and agriculture. Retreatment of wastewater is permitted, and infrastructure is specified including conveyance lines, wastewater treatment plants and desalination facilities - such infrastructure being either actual or potential (Ibid., 187)'.

Thus WAS runs allow for the determination of the optimal usage of water resources.

Fisher *et al.* make it clear that 'optimal' does not mean 'market optimal'. For Fisher *et al.* domestic water markets do not allow the best allocation of water resources. Indeed, for them, the value of water should not be limited to its pure private value. In fact their model takes into account social values and policies. This is done by modifying the position of the demand curve for water in the WAS model. Fisher *et al.* present a model that takes as given Israel's necessity to

---

[1] The figure can be compared with Israel's and West Bank and Gaza's GDP. According to the World Bank, in 2005, Israel's GDP was $123,434 million. West Bank and Gaza's GDP was $3,454 million. In the case of the West Bank and Gaza the cost of water in dispute would thus be far higher than for Israel with regard to their respective GDP. For West Bank and Gaza, $100 million represent 3% of its GDP.

subsidize agriculture and therefore develops a model enabling the conduct of such policy toward agriculture.

Fisher *et al.*'s 'new approach' to the Middle East does not stop there. Fisher *et al.* explain that the question of water ownership should not be overemphasized. Applying the Coase Theorem to water resources allocation, they intend to demonstrate that water ownership is 'just' a 'property right entitling the owner to the economic value of the water. A dispute over water ownership can be translated into a dispute over the right to monetary compensation for the water involved (Ibid., 63)'. For them, water ownership and water usage are analytically independent. The initial allocation of ownership rights is neutral: it does not affect the efficient usage of water resources. They explain further:

> 'All water users are buyers, in effect - whether they own the water themselves or purchase it from another party. An entity that owns its water resources and uses them itself incurs an opportunity cost equal to the amount of money it could otherwise have earned through selling the water. An owner will use a given amount of its water if and only if it values that use at least as much as the money to be gained from selling. The decision of such an owner does not differ from that of an entity that does not own its water and must consider buying needed quantities of water : the non owner will decide to buy if and only if it values the water at least as much as the money involved in the purchase. Ownership only determines who receives the money (or the equivalent compensation) that the water represents (Ibid., 63)'.

There is a double argument here. First, recalling the Coase Theorem, Fisher *et al.* argue that the initial allocation of water rights is neutral from an efficiency point of view. Second, by expressing water ownership in monetary terms, they suggest that property rights in water could be traded-off for relatively small sums. In any case, the double argument is used to 'deemphasize' the question of property rights allocation.

However, recognizing that water remains 'an emotional topic' and that it might be difficult for countries to negotiate over water ownership rights, they go further into their argumentation showing that the gains from 'cooperation'- that is trade in water use permits - are greater than the gains from any change in ownership rights. They conclude:

> 'The lesson is clear: ownership is surely a symbolically important issue, and symbols really matter, but cooperation in water reduces the practical importance of ownership allocations - already not very big -to an issue of very minor proportions'(Ibid., 208).

This demonstration is done through runs of the WAS model. Runs of WAS allow determining the benefits that each party gains from 'cooperation' – that is, participation in a system of tradable rights. These benefits are then compared to the benefits stemming from additional water ownership. Two sets of results are presented. In one case, runs are made for different assumptions about water ownership of the Mountain Aquifer between Israel and Palestine. In the other case, the different divisions of ownership of the Jordan River between Israel and Jordan are examined[2].

The gains resulting from a change of ownership allocation are calculated with and without 'cooperation'. The gains from an ownership change are higher without 'cooperation' than gains from an ownership change with 'cooperation' but they are in both cases much less than the gains from 'cooperation' – i.e. trade in water permits. The gains from 'cooperation' are almost never zero. They would be zero if the allocation of ownership rights would be optimal. However, because population and other factors change, this allocation cannot remain optimal across time. Fisher *et al.* demonstrate thus that the gains from 'cooperation' (i.e. trade in water rights) are always greater than those emerging from a change in ownership rights. And in both cases, it seems that the status quo regarding the ownership allocation is favoured[3]. They conclude: 'again we see that the gains from cooperation are bigger than those from shifts in ownership' (Ibid., 213).

Finally, Fisher *et al.* present the results of adding a third partner to bilateral cooperative arrangements (that is either Jordan joins an Israel-Palestine agreement or Palestine joins an Israel-Jordan agreement). They explain, 'in all cases, Israel gains and its existing partner loses when the third party enters the arrangement' (Ibid., 216).

> 'With the assumed ownership divisions, Israel has a great deal of water and acts as a net water permit seller in its cooperative arrangements. Palestine and Jordan, on the other hand, are net water permit buyers. The entry of either into an arrangement between Israel and the other necessarily increases the demand for water permits from Israel and raises the price at which such permits get sold. This benefits the net seller but disadvantages the net buyer. Fortunately it is always possible

[2] Current division of the Mountain Aquifer: 76% to Israel, 24% to Palestine. Current division of the Jordan River : 92% to Israel, 8% to Jordan (662 million m^3 to Israel, 55 million m^3 to Jordan).

[3] Fisher *et al.'s* proposal could raise important opposition. The 2009 Geneva Initiative's Water Annex, which represents the results of negotiations between Israeli and Palestinian civil society groups, recognizes the necessity of a 'just and rightful re-division of water' according to international water law principles (article 3). For many people in Israel and Palestine, the re-division of water is a necessary step toward a peace agreement as well as the creation of a joint water commission (cf. footnote 4). This re-division should include the Mountain Aquifer, the Coastal Aquifer and the lower Jordan River. 'Priority should be given to meeting vital human needs'. In the 2009 model water agreement, the re-division of water is accompanied by the commitment not to cause significant harm to the other Party (article 4)- a way to address the Palestinian wastewater and pollution mismanagement.

to redistribute the gains so that everybody benefits from increased cooperation' (Ibid., 217).

To adopt an economic approach and to value water in a dispute can certainly help resolve water conflicts, but Fisher *et al.*'s arguments have many weaknesses.

The Problem of Water Policies' Harmonization

From an economic point of view, Fisher *et al.*'s analysis of the impact of Israeli agricultural subsidies on the Palestinians presents little in the way of convincing results. In the case of an agreement between Palestine and Israel on trade in water permits, Fisher *et al.* argue that 'the effects on Israel or Palestine of an agricultural subsidy by the other are either negligible or, because of the international trade effect, slightly positive' (Fisher *et al.* 2005, 70). They explain that it is the party doing the subsidizing that bears the cost due to the increase in water demand and thus the increase of the scarcity value of water resources. This might be true indeed, but in the case of Israeli agricultural water subsidies, the Palestinian consumers will also be concerned by the increase in the price of water resources. Besides, if there is trade in water permit without harmonization in water policies, the Israeli agricultural products will be sold at a lower price than Palestinian products, thusly creating unfair competition.

However, Fisher *et al.* reject these arguments, explaining that an agreement on trade in water permits between Israel and Palestine would not require an agreement on domestic water policies. 'The right of each country to set its own national policies toward water should not be questioned' (Ibid., 69).

More generally, from a point of view of water use efficiency, Fisher *et al.*'s strong defence of Israeli agriculture can be subject to critique. If the region suffers from a water shortage, it is primarily because of the agricultural use of water resources (agriculture uses more than 60% of water resources). In order to reduce the pressures on water in the region, it would be rational to first reduce water consumption in the agricultural sector. Thus, before proposing trade in water permits across borders, it would be more useful to create domestic water markets enabling the reallocation of water resources among the different domestic sectors.

Another way of coping with the water crisis in the region would be to increase reliance upon the importation of 'virtual water', that is, water used in the production process of crops and other agricultural or industrial products (Allan 1997). Of course, such a solution clearly leaves aside the security context of the Middle East. Without peace in the region, Israel's policy toward agriculture is unlikely to evolve. Berck and Lipow (1994), looking at the link between agriculture and water use in the region, clearly note 'Israeli water policy is driven by the perceived necessities of warfare and ideology, rather than by concern for economic efficiency' (Berck and Lipow 1994, 291).

Coase Theorem and Water Resources

A key point in Fisher *et al.*'s analysis is the opposition between 'cooperation' and 'ownership' (see Fisher *et al.* 2005 '*cooperation versus ownership*', 197). In their book, 'cooperation' means trade in water permits between Israel, Palestine and, potentially, Jordan. 'Ownership' means ownership of a certain quantity of disputed waters (the Jordan River and the Mountain Aquifer).

By measuring and comparing the gains from cooperation (i.e trade in water permits) with the gains resulting from a change of ownership in allocated water quantities (with and without trade), Fisher *et al.* want to show that water ownership allocation is a secondary matter in comparison with the way water is *used.* Although this is a slightly modified version of the Coase Theorem, Fisher *et al*'s theoretical demonstration in the fourth chapter clearly rests on the former theorem (Ibid., note 3 p 63).

The application of the Coase Theorem to water allocation problems in the Middle East requires, however, a lot of prudence. In his 1960 article, *the Problem of Social Cost*, from which the 'theorem' was later derived, Coase's main concern was how to cope with harmful effects of firms. Denouncing the arbitrary decisions of British and American Courts in liability cases, Coase shows the reciprocal nature of some particular harmful effects (such as smoke, noise, odours, etc.). He notices that there are usually no clear rules determining who is allowed to do what. From an efficiency point of view, what matters is determining whether the gain from preventing the harm is greater than the loss that would be suffered if the action that produces the harm is stopped. As it is impossible to know this in advance, what is needed is a clear definition of who has the right to do what, and then let the individuals bargain freely. The initial allocation of property rights is neutral in the absence of transaction costs, meaning that the result of the bargain is independent from the initial delimitation of rights from an efficiency point of view (if there are no transaction costs). It is interesting to note that Coase makes it clear that the absence of transaction costs is 'an unrealistic assumption' (Coase 1960, 15) and recognizes the importance of government regulations in certain cases involving particular types of harmful effects:

> '[. . .] there is no reason why, on occasion, such governmental administrative regulation should not lead to an improvement in economic efficiency. This would seem particularly likely when, as is normally the case with the smoke nuisance, a

> large number of people are involved and in which therefore the costs of handling the problem through the market or the firm may be high' (Ibid., 18).

In this sense, an essential part of Coase's demonstration is to show the importance of the institutional framework and law in the search for economic efficiency.

Fisher *et al.* do not seem to address the question of transaction costs in their application of the Coase Theorem to water resources. In the particular case of water in the Middle East, transaction costs - including information, bargaining and enforcement costs - are far from being non-existent. The political and security context in this particular case is very different and much more complex than the domestic context described by Coase in his demonstration. The allocation of ownership rights is not a secondary matter in comparison to water uses. In the particular context of the Middle East, trade in water use rights will occur with high transaction costs, and, for this reason, the initial allocation of ownership rights is a determining factor.

Besides, it could be argued that wealth effects and distortions of the bargaining powers are even bigger in the case of water resources than in other cases of harmful effects. The allocation of a right authorizing or protecting against the creation of a harmful effect such as noise, pollution or odours is a clear reciprocal matter, whereas in the case of the allocation of ownership of a resource already existing in nature, like water, the reciprocal character of the allocation is far less evident.

The Coase Theorem is more easily applicable to the allocation of water pollution permits than for the allocation of water ownership. In fact, the reciprocal nature of the allocation of water resources depends on the institutional framework: that is, *what* exactly is allocated. If it is ownership rights, there is no reciprocity because you have the water or you do not have the water. If it is water use rights, with the use being clearly delimited and regulated according to a water plan, the reciprocal character of the allocation is higher than in the case of ownership allocation. Once again, it clearly appears that the institutional and regulatory dimensions are central aspects of Coase's work. Fisher *et al.* unfortunately seem to neglect this fundamental aspect in their demonstration. In other words, the problem of the allocation of water could be resolved by accepting the common ownership of the disputed water resources and allocating to the different parties water use rights, and not ownership rights, as proposed by Fisher *et al.*

Recalling that the allocation of property rights is a secondary matter in comparison to water use, Fisher *et al.* make an interesting proposal that was regrettably not further developed. They explain that trade in water permits could begin even before the question of property rights' allocation is settled by making payments into an interim escrow fund. They explain, 'that fund would be jointly managed and provide a source of financing for mutually desirable infrastructure. Negotiations over property rights would effectively become negotiations over shares of or obligations to the fund plus entitlements to future payments. This is as it should be, since water property rights are a matter of money' (Ibid., 68).

This is indeed an interesting solution but it also raises many questions, such as: which water will be concerned by the escrow fund? How will the interim escrow fund deal with the existing allocation of water ownership? Some actors might use water without making payment to the escrow fund arguing that the water they use is already theirs.

We suppose that Fisher *et al.* proposal concerns all the disputed waters and, in this sense, any party willing to use the water would have to make a payment to the fund. If this is the case, it might trigger important changes in the current allocation of water resources. Even if this fund is only temporary, it will require negotiations over a common water plan providing at least some water for the environment, and determining which quantity of water is available for what type of use, consumptive or non-consumptive.

Also important to ask are, what will happen once the escrow fund program is terminated? Will there be a common management of the trading permit program?

In fact, the overall analysis of Fisher *et al.* lacks institutional and regulatory approaches. The dissociation between *ownership* and *use* rights is not sufficient to resolve the water allocation problem. Any agreement between the parties should first define the exact nature of the property regime of the disputed resources, not just who gets what. Here, 'property regime' is taken in a broad sense. It does not mean who gets what, but how and who will make the decision over which resources. In the case of cross border aquifers, such as the Mountain Aquifer, and international rivers such as the Jordan, the property regime of such water resources would normally involve the creation of a permanent, common decision-making body[4]. The definition of the disputed water resources as common property is an essential preliminary step before the introduction of a market in water permits. It is also the only way to truly deemphasize the question of ownership allocation. If ownership is common, then only water use is allocated. Before introducing a trading scheme between the parties, it is necessary to agree on an institutional

[4] Interestingly the Water Annex released in 2009 by the Geneva Initiative – Israeli and Palestinian civil society groups working on a model of Israel-Palestinian Peace Agreement- provides for the creation of a 'joint water commission' composed of three representatives from each side and an elected 'neutral chairman' who shall be 'of another nationality' and a 'qualified water professional' (article 6).

framework to regulate the water market. It is thus necessary to primarily define the common property of the resources. That is, to answer to the following questions: What would be the rights and duties of a permit holder? What would be the scope of the market? Who would be responsible for the monitoring of water transfers and uses? How would disputes over water transfers be settled? How would the allocation for the environment be protected? Which rules will protect affected third parties? What should be the provisions for the areas of origin? Clearly, to say that water permits represent water use permits alone is not sufficient to resolve these issues. These uses need to be regulated.

It is uncertain whether Fisher *et al.*'s 'new approach' will contribute to the 'de-emotionalization' of the water dispute in the Middle East. Their analysis lets pass aside too many questions, most notably institutional considerations. As we tried to show, Fisher *et al.*'s understanding of property rights and ownership does not allow any discussion on institutional frameworks and decision-making procedures essential for the establishment of a trading scheme and a true resolution of the water dispute. In the case of cross-border water resources, property rights' allocation should not mean who gets what, but rather how will these resources be managed, who will make decisions on what resources, through which procedures, and following which principles. To define a property regime for water resources means defining a web of interests including the hub, scope and the architecture of the web- meaning the decision-making procedures and its actors-, but also which interests are legitimate, and which principles govern their ordering.

Despite these weaknesses, Fisher *et al.*'s analysis has the merit of shedding light on the problem of many international agreements, which lack flexible mechanisms for adaptive water management. The introduction of trading schemes is one mean among others to introduce flexibility into the management of international watercourses. In addition, if these market-based systems are to be equitable in the long term, they have to be accompanied by an institutional framework providing not only for operational and decision-making rules, but also for constitutional-level rules (Schlager and Ostrom 1992, Bruns 2006). Because conditions change over time, it is important that international water agreements provide 'rules for making rules,' that is, constitutional rules according to the three levels' definition of a property regime by Schlager and Ostrom. Thus, water rights' allocation should not be permanent, and the rules attached to these rights should be modifiable.

As we saw with Fisher's proposal, international water markets are no easy solution. The introduction of trading schemes across borders raises fundamental questions. And, before such schemes are introduced, many issues need to be settled, such as, at a minimum, the creation of a well-functioning common decision-making body.

The introduction of inter-state trading schemes has recently been tested in two large river basins in the US and in Australia. The Colorado River Basin and the Murray-Darling Basin are two cases that can bring interesting insights into the discussion over the introduction of market-based solutions in international river basins. We are aware that these two cases are federal and not international cases. However, we argue that in the same manner as federal states played an important role in the development of international water law principles, they might today create important precedents for the way international watercourses are managed.

Chapter 7

Colorado River Basin Case Study

Drought in the Southwest of the United States has become the rule rather than the exception. Concerns about the diminishing levels of the Colorado River flow are real. In September 2007, the storage in Colorado River reservoirs was at only about 54% of its capacity. It is estimated that the Colorado River Basin is experiencing its worst drought in 100 years. In fall 2007, Southern California was ravaged by historic wildfires. Various scientific reports have demonstrated the dramatic effects of climate change in the Colorado River Basin (IPCC 2001, US Geological Survey 2004).

In this context, the seven basin states and the Secretary of the Interior have been negotiating for two years to find an agreement on a way to manage the Colorado River in case of continued shortage. Indeed, whereas there were guidelines to operate the river allocation in normal and surplus years there were no guidelines in case of shortage. With sustained drought in the Colorado River Basin, any further water demand will have to be satisfied either by conservation or by water trading ('augmentation projects' such as weather modification are also studied as potential solutions). We retrace here the developments of the last decade in the Colorado River Basin and the different attempts to introduce some flexibility into the Law of the River. We give here one interpretation of the Law of the River; we are aware, however, that its interpretation is subject to many controversies among

ISBN: 9781843393610. Published by IWA Publishing, London, UK.

the seven basin states. At issue here, for example, is whether interstate water transfers are compatible with the logic of the 1922 Compact and, in that case, under which conditions. What kind of modifications would that require? How would an interstate water market in the Colorado River Basin look like? What would be the legal, environmental, social consequences of such an evolution?

Because these issues were analysed for the adoption of shortage guidelines in the Environmental Impact Statement that the Bureau of Reclamation released in November 2007, we believe the Colorado River Basin experience can bring interesting insight into the discussion over market-based solutions along international watercourses.

GENERAL OVERVIEW OF THE COLORADO RIVER BASIN

The Colorado River is born at about 10,000 feet (2,800 m) in the Rocky Mountains of Colorado and flows southwest for 1,470 miles (2,330 km) to the Gulf of California in Mexico (the natural course of the river flows into the Gulf of California, but the heavy use of the river as an irrigation source for the Imperial Valley has desiccated the lower course of the river in Mexico such that it no longer consistently reaches the sea). It is the international boundary for 17 miles (27 km) between Arizona and Mexico. The 1,360 miles (2,188 km) of its route in the United States makes it the nation's fifth longest river. It drains a large portion of the North American continent covering 242,000 square miles (629,100 km^2) in the United States and 3,000 square miles (4,828 km^2) in Mexico. The drainage basin comprises about one-twelfth of the area of the continental United States. Despite the size of its drainage area, the Colorado River ranks only sixth among the nation's rivers in volume of flow with an average annual flow of 15 million acre feet at Lee Ferry, the Colorado River Compact division point. However, in the Lee's Ferry record, flow in the recent years is the lowest in more than 100 years of observations, averaging only 5.4 m.a.f. during the 3-year period 2001-2003. The lowest previous 3-year average flows were 7.3 m.a.f. from 1954 through 1956 and 8.0 m.a.f. from 1933 through 1935 (US Geological Survey 2004).

The Colorado and its tributaries drain South-western Wyoming and western Colorado, parts of Utah, Nevada, New Mexico and California, and almost all of Arizona (see map below). Three fourths of the basin is federal land devoted to national forests and parks and Indian reservations. The Lower Colorado River separates two great deserts, the Mojave on the California (Western) side and the Sonoran on the Arizona (Eastern) side.

The Colorado River drainage area receives on average about 14 inches (35cm) of precipitation per year and can truly be called the 'Great American Desert'. Basin water supply is strongly dependent on snowmelt runoff in high elevation portions of the Upper Basin, with about 15% of the watershed area producing about 85% of the entire basin's average annual runoff (Department of Water Resources of California 2005).

Colorado water is used for irrigation purposes, hydroelectricity production and municipal and industrial supply. About two-thirds of the water flowing in the Colorado River and its tributaries is used for irrigation (2 million acres of irrigated land). Around twenty dams were built for hydroelectric, storage and regulation purposes on the river. As a result, the river has more than 60 million acre-feet of storage capacity (about four years of average flow of the Colorado River) and 4,000 megawatts of hydro-electric generating capacity. The first major development began in 1928 when Congress passed the Boulder Canyon Project Act authorizing construction of Boulder (now Hoover) Dam creating Lake Mead. In the mid-1960s, Glen Canyon Dam was completed, impounding Lake Powell. Many additional projects have been completed since then. The Colorado River provides municipal and industrial water for more than 24 million people living in the major metropolitan areas of Los Angeles, Phoenix, Albuquerque, Las Vegas, Salt Lake City, Denver, San Diego and hundreds of other communities in the seven states (Anderson 2002).

THE 'LAW OF THE RIVER'

The Colorado River Divided in Two Sub-basins

In the USA, interstate water allocation is essentially done through bargaining among the various riparian states. However, in contrast to treaties on international watercourses, compacts among US riparian states are ratified by Congress and signed by the President. As presented in the second chapter, the US Supreme Court also plays a major role in interstate water allocation. Suits dealing with interstate allocation before the Supreme Court do not involve only states, but private parties as well.

The 'Law of the River' – i.e. the body of rules governing the Colorado River - rests essentially on the 1922 Compact, but also on state and federal laws, court decisions and international treaties.

The 1922 Colorado Compact is the first-ever river compact in US history. Concerned by the rapid development of California, and aware of the consequences of the Western rule of prior appropriation on interstate waters (*Wyoming v. Colorado*, 1922, 259 US 419), upstream states proposed to negotiate a compact

Figure 5 Colorado River Basin in the United Sates (Credit to Colorado River Commission of Nevada).

to define individual states' right to the river water. The basis to determine each state's share was to be the amount of irrigable land within a state. Because at the time of the negotiations no good data was available and because it was difficult anyway to agree on an allocation for each state, it was decided that the Colorado apportionment be done for two sub-basins instead than for each individual states. The agreement provides that the Upper Basin (Wyoming, Colorado, Utah and New Mexico) like the Lower Basin (California, Arizona and Nevada) receives 7.5 million acre feet per year each. The Compact mentions also that the Lower Basin could increase its apportionment by one million acre feet per year. The Upper Basin states are required to guarantee an aggregate of 75 million acre feet per period of ten consecutive years to Lower Basin states. This apportionment was done 'in perpetuity'. The problem, however, is that this apportionment was based on an estimate of Colorado annual flow of 16.4 million acre feet, whereas it is today closer to about 13.5 million acre feet.

State Water Allocation

Arizona refused to sign the Compact because the apportionments were not made directly to the states but to the sub-basins. Arizona wanted an apportionment per state in order to limit the application of the law of prior appropriation. An apportionment per basin meant that Arizona had to compete with the fast-growing California for prior uses within the Lower Basin. For that matter, Arizona searched to develop its use of the Colorado. In the early 1940s, the Central Arizona Project was presented as a good way to effectively use its apportionment while delivering water to the growing population in the south-central part of the state. The federal government accepted to support such a project if Arizona ratified the 1922 Compact, which Arizona did a little later on. But, such a project supported by the Department of the Interior could not be implemented before California and Arizona resolved their dispute over the Colorado River. Unable to come to an agreement amongst them, Arizona asked the US Supreme Court in 1952 for a judicial apportionment. More than ten years later, the decision *Arizona v. California* held that by passing the Boulder Canyon Project Act in 1928, Congress had, in effect, apportioned the mainstream of the Colorado River with California receiving 4.4 million acre-feet, Arizona 2.8 million acre-feet, and Nevada 0.3 million acre-feet. Moreover, the decision interpreted the Boulder Canyon Project Act as empowering the Secretary of the Interior to act as watermaster of the Lower Colorado River, to apportion future surpluses and shortages among the states and even among users within the states.

By contrast, the Upper Basin states managed to negotiate a compact to apportion their share of the Colorado water. The 1948 Compact apportions water in percentages of the Upper Colorado water assigning 51.75 % to Colorado, 23% to Utah, 14% to Wyoming and 11.25% to New Mexico.

On top of these various apportionments, the Colorado Basin states have an obligation under the 1944 Treaty between Mexico and the USA to provide Mexico with a minimum of 1.5 million acre-feet. The following table summarizes Colorado River apportionment.

States		Million acre-feet per year (m.a.f.)
Upper Division States		**7.5***
Colorado	3.9	
Utah	1.7	
Wyoming	1	
New Mexico	0.85	
Lower Division States		**7.5**
California	4.4	
Arizona	2.8	
Nevada	0.3	
Mexico		**1.5**
Total		**16.5**

*the portion of Arizona that lies within the Upper Colorado Basin was also apportioned 50,000 acre-feet annually in the 1948 Compact (7,45+0,05 (Arizona)=7,5).

Table 1 Colorado River Apportionment.

TESTING THE LIMITS OF THE LAW OF THE RIVER

For the first time, in 1990, the three Lower Basin states exceeded their 7.5 m.a.f. allotment. This situation triggered off a series of negotiations among the Colorado Basin states and the Secretary of the Interior. Different proposals were made to introduce flexibility into the allocation system and increase efficiency of water use. In 1999, the Department of the Interior adopted a final rule providing a framework for interstate water banking. More recently, in December 2007, the Secretary of the Interior signed an important record of decision opening new avenues for water management in the Colorado River Basin, essentially by allowing the storage of 'intentionally created surplus' in Lake Mead.

Private Proposals of Interstate Water Transfers

In the mid-1980s and early 1990s, different private groups proposed to sell or lease Colorado water to California. One of the most important projects was designed by the Galloway Group, a corporation based in Colorado. In August 1984, the San Diego County Water Authority paid Galloway $10,000 for an option to lease 300,000-500,000 a.f. per year for forty years (Gross 1985, 937). At that time, San Diego was receiving more than its formal Colorado River entitlement. But, aware of the growing consumption of the other Lower Basin states and of its fifth-ranking priority out of the seven Californian users (according to the 1931 Seven Party Agreement), San Diego was looking for alternative water supplies. When the agreement was negotiated, San Diego was using approximately 269,000 a.f. of Colorado River water whereas its formal entitlement was 112,000 a.f. (Gross 1985, 937).

The project was never carried out. The Upper Colorado River Basin Commission opposed the project. The issue was then whether the 1922 Compact, by apportioning Colorado water between the Upper and Lower Basin, established a geographical limit to water use or whether Colorado water could be considered an article of commerce that could be freely transferred across state lines.

The Upper Basin states found that the Compact expressly prevents trans-basin transfers. Article VIII of the Compact mentions that all rights to beneficial use under the Compact 'be satisfied solely from the water apportioned to that Basin in which they are situated'. Besides, article III (e) states that 'The states of the Upper Division shall not withhold water, and the states of the Lower Division shall not require the delivery of water, which cannot reasonably be applied to domestic and agricultural uses'. Upper Basin states, recalling the spirit of the 1922 Compact, explained that the Compact apportioned 'uses' and not 'ownership' of the Colorado River. Article I of the Compact mentions, indeed, 'the major purposes of this Compact are to provide for the equitable division and apportionment of the *use* of the waters of the Colorado River System' (emphasis added). Thus unused Colorado water apportionment could not be sold to another state. In this sense, commercial use (that is water sale to another state) was not considered as being a 'beneficial use' for the selling state.

Aware of its limited chance to win a case invoking the commerce clause, the Galloway Group did not try to bring the case before the Court. Indeed, as already studied in the former chapter, in *Sporhase v. Nebraska*, the Court's recognition of water as an article of commerce is limited by a series of factors including interstate compacts. In his opinion, Justice Stevens explains 'the legal expectation that under certain circumstances each state may restrict water within its borders has been fostered over the years not only by our equitable apportionment decrees, see, e. g., *Wyoming v. Colorado*, 353 U.S. 953 (1957), but also by the negotiation and enforcement of interstate compacts. '*Our law therefore has recognized the relevance of state boundaries in the allocation of scarce water resources'* (*Sporhase v. Nebraska*, emphasis added).

Finally, Colorado statutes represented another obstacle to Galloway's Project. Adopted after the Court's ruling in *Sporhase* and designed to face attack under the commerce clause, Colorado out-of-state water export statutes are closely tied to the Colorado Compacts. After 1982, many states redesigned their water export statutes in such a way to comply with the new ruling. Different solutions were found to limit or prohibit out-of-state water exports. Many states, for example, require legislative approval for an interstate water transfer to take place. One also found legislation restricting water transfers to intra-basin transfers only. Although not entirely prohibiting interstate transfers, such legislations represent anyway an important limitation to such projects. Moreover, some states directly ask Congress to approve their water export prohibition. This is usually done through congressional ratification of an interstate compact (for example, the Yellowstone River Compact containing a provision prohibiting interbasin export without the consent of all the state parties). In the Water Resources Development Act of 1986, the Congress prohibits the exportation of water outside the Great Lakes Basin without the consent of all the governors of the Great Lakes states. Aware that the 1922 Compact imposed restrictions on transbasin transfers, Colorado decided to tighten its statutes to align with the requirement of the Compact. Thus, the statutes stipulate that

> 'Prior to approving an application, the State Engineer, Ground Water Commission or Water Judge, must find that: (a) the proposed use of water outside this state is expressly authorized by interstate compact or credited as a delivery to another state pursuant to section 37-81-103 or that the proposed use of water does not impair the ability of this state to comply with its obligations under any judicial decree or interstate compact which apportions water between this state and any other state or states, [. . .] (c) the proposed use of water will not deprive the citizens of this state of the beneficial use of waters apportioned to Colorado by interstate compact or judicial decree' (Colorado Statutes §37- article 81–101, 1984).

The request of consistency with the interstate compacts clearly limits the possibility of water exports from Colorado.

Another project is also important to present for our analysis of the challenges facing the Law of the River. In 1990, the Resource Conservation Group, Inc. (RCG), a partnership composed of a corporation, RCG and various investors based in Colorado, proposed to lease Colorado River water to San Diego. Aware of the Galloway Group's failure to obtain an approval of the Upper Basin Commission, they developed a more sophisticated argument. They basically distinguished between Upper Basin 'surplus' water (i.e. above 7.5 m.a.f.) and Upper Basin allocated water. According to them, article III(e) of the Colorado Compact -

stating that water which is not beneficially used in the Upper Basin cannot be withheld from the Lower Basin- concerned only 'surplus' water, that is water above 7.5 m.a.f. For them, thus, the Upper Basin states have an absolute right to 7.5 m.a.f. per year and can store the excess water which they do not use. Contrary to 'surplus' water, Lower Basin states do not have a legal right to the free use of this water. The RCG argument goes even further by suggesting that unused but apportioned water is similar to water actually used in the Upper Basin. They propose to lease water to San Diego, water that would be made available by private owners of water rights. RCG would take the farmer or rancher's water right on a rotational basis, paying for the use but not taking his right. The project was rejected essentially on the ground that the Colorado Compact provided territorial limitation to water use. Notably, article III (a) mentions that the Upper and Lower Basin are apportioned, respectively, the '*exclusive* beneficial consumptive use of 7.5 m.a.f. per annum' (emphasis added).

Although never undertaken, RCG's project is a serious attempt to test the limits of the Law of the River. It raised delicate questions regarding the interpretation of the Compact, notably the status of unused but apportioned water (unused apportionment), still debated today.

The most recent transfer proposal is the Roan Creek Project which was abandoned in 1993. Chevron Oil, located in DeBeque, Colorado, wanted to lease its water rights issued for oil shale production to Las Vegas, Nevada.

Regularly, Colorado had to justify its opposition to water transfer proposals into Lower Basin states. The Colorado River Water Conservancy District, the Colorado state water agency, explains in its policy statement that:

> 'Under most interpretations of the Compacts, the Upper Basin states do not have a clearly quantified allocation. Therefore, one Upper Basin state selling a portion of its unquantified entitlement is problematic, at best. At worst, it introduces Lower Basin interests into any eventual resolution of ambiguities in the 1948 Upper Basin Compact and changes allegiances within the Upper Basin when negotiating ambiguities in the 1922 Compact'[1].

These interstate water transfer proposals underline the growing demand for Colorado River water in some parts of the Basin and the need to create new rules to govern the reallocation of water.

They also bring to light the lack of institutional framework among the Colorado River Basin states to deal with interstate water marketing proposals.

---

[1] http://www.crwcd.org/media/uploads/policy-interstate_marketing.pdf last visited in February 2010.

California's Interstate Water Bank Proposal and the Beginning of the Negotiations over Interstate Transfers in the Colorado River Basin

As already mentioned, in 1990, the Lower Basin states' consumption exceeded their 7.5 m.a.f. allotment for the first time. In 1991, they requested to the Secretary of the Interior the delivery of an amount of water exceeding their 7.5 m.a.f. allotment. An arrangement was reached between Upper and Lower Basin states, but Colorado Governor Romer conditioned his agreement on the commitment of California to start a process to reduce its consumption to its 4.4 m.a.f. allocation. Indeed, for decades, California exceeded its 4.4 m.a.f. apportionment, using around 5.2 m.a.f. This was tolerated as the two other Lower Basin states, Nevada and Arizona did not use their full allotment. However, with the rapid growth of Las Vegas in the late 1980s and early 1990s, and the near completion of the Central Arizona Project, Nevada and Arizona came closer to the full utilization of their Colorado River allocation, inevitably leading to the Lower Basin exceeding the 7.5 m.a.f. provided in the 1922 Compact.

In this context, it soon became obvious that the Law of the River had to be revisited in order to allow some kind of interstate transfers among Colorado River Basin states. Various proposals introducing flexible mechanisms to reallocate water were made. From the 1990s until today, one observes a competition for authority in the Basin between the Bureau of Reclamation and the states in designing the rules of interstate transfers. Thus, the procedures to establish the institutional framework of such transfers were lengthy and complicated. In the next section we present the development of the negotiations over interstate transfers along the Colorado River.

In 1991, the Colorado River Board of California proposed a 'Conceptual Approach for Reaching Basin States Agreement on Interim Operation of the Colorado River System Reservoirs, California's Use of Colorado River Water above Its Basic Apportionment, and Implementation of an Interstate Water Bank.' Thus, the Conceptual Approach included the proposal for the creation of an interstate water bank. The proposal provided that historically consumptively used water could be transferred to users in other states for one year. Water would be purchased by the bank at a uniform price fixed by the seven states. Water would then be made available for sale to users in the seven states according to a formula allocating that water according to the state's share of the basic Colorado River apportionment (MacDonnell 1994).

California's proposal was rejected, especially by the state of Colorado, which argued that California needed to reduce its use to 4.4 m.a.f. and pushed for further reallocation within California before enabling interstate transfers.

California's proposal started a process of modernization of the 'Law of the River' and set off a series of reflections on the introduction of market mechanisms along the Colorado River Basin. Thus, soon after the Californian proposal, the Bureau of Reclamation issued a discussion paper proposing the creation of a Lower Basin Water Bank. Unused but apportioned Lower Basin water could be stored in a bank (up to 4.5 m.a.f.). The three Lower Basin states would have access to this stored water proportionally to their share of the Lower Basin apportionment (MacDonnell 1994, 25).

At issue in these discussions is the status of apportioned but unused Colorado water. California's proposal and the Bureau of Reclamation's proposal outline very different characteristics. California's proposal concerns the entire Basin whereas the Bureau of Reclamation's proposal is limited to reallocation among the Lower Basin states. Besides the scale, the actors are also different. In the Californian case, users would be allowed to buy and sell water to the Bank whereas the Bureau of Reclamation proposes interstate transfers operated by the states.

A third issue soon arose in the discussion over interstate transfers along the Colorado River: what role for the Indian Tribes? Denouncing the absence of any reference to them in the proposals, the Indian Tribes created the 'Colorado River Tribal Partnership'[2] and issued a position paper intending to enter the negotiations over interstate transfers along the Colorado Basin. A major concern for the tribes was the protection of their unused entitlements. They thus proposed to lease this water for off-reservation use. They argued that such a lease should be approved by the seven states, the tribal partnership and the Secretary of the Interior (MacDonnell 1994, 26). By requesting this series of approvals, the Indian Tribes looked for the recognition of their apportionment and their securitization.

In 1994–1995, negotiations accelerated. Aware that the Bureau of Reclamation was preparing a regulation in which it would have a central role in the management of Colorado River transfers, the Lower Basin states, especially Arizona hurried up to develop their own strategy.

In May 1994, the Bureau of Reclamation released some proposed regulations, including the creation of a federally-operated water bank utilizing storage in Lake Mead. It also reinforced its authority to control water use of Colorado River water. The proposed regulations provided for the interstate lease of all or part of an entitlement for a period limited to 50 years. Only water saved through extraordinary

[2] The Ten Tribe Partnership is composed of ten Indian Tribes : Chemehuevi Indian Tribe, Cocopah Indian Tribe, Colorado River Indian Tribe, Fort Mojave Indian Tribe, Jicarilla Indian Tribe, Navaho Nation, Quechan Tribe, Northern Ute Indian Tribe, Southern Ute Indian Tribe and Ute Mountain Indian Tribe.

conservation measures or land fallowing could be transferred. Thus, the regulations did not address apportioned but unused water (unused apportionment). The Bureau of Reclamation instead created another category of water: 'conserved water', defined as 'water within the limit of an entitlement which has been put to historical beneficial use and saved as a result of specific conservation measures which are identifiable, quantifiable and verifiable'[3]. The regulations also provided for the establishment of an evaporation tax against banked water. In contrast with its former proposal, this Reclamation's proposal directly concerned holders of entitlements and not the states. It dealt with 'conserved' water only, and not with apportioned but unused state allocation (unused apportionment).

Two Lower Basin states reacted quickly to these proposed regulations. Nevada proposed a water bank operated by a commission comprised of the three Lower Basin states, each with an equal voting power. Arizona went along with its own water bank.

In its 1999 annual report, the Arizona Water Banking Authority (AWBA) recalled its major concerns regarding the 1994 Bureau of Reclamation's proposal. Primarily, they refused that the interstate banking system be controlled by the Reclamation. More precisely, they rejected the regulations granting the Secretary of the Interior the authority to reduce entitlements on the basis of non-use and to reallocate water within the states. According to them, the proposal was in violation of *Arizona v. California*. They rejected the proposal because 'it put Arizona's long-term supply of Colorado River water at risk' (AWBA 2000).

The 1999 AWBA report mentioned that the rejection of the 1994 proposal marked the end of the process driven by federal rulemaking, and, indeed, from that date on the initiative went back to the states, and particularly to Arizona which adopted its own water bank.

INTERSTATE WATER BANKING

Arizona Water Banking Authority

In fall 1995, the President of the Arizona State Senate and the Speaker of the Arizona House of Representatives created the Joint Legislative Committee on Colorado River Water Issues to examine ways to increase diversions and use of Arizona's allocation. House Bill 2494, which created the Arizona Water Banking

[3] US Bureau of Reclamation, Proposed Regulations for Administering Entitlements to Colorado River Water in the Lower Colorado River Basin, May 6, 1994 (the regulations were not published in the Federal Register). Quoted in MacDonnell, 1995, p28.

Authority (AWBA) passed the legislature and was signed by Governor Fife Symington on April 30, 1996.

Contrary to former proposals, Arizona's bank plan allowed Arizona to have an exclusive authority on its Colorado water entitlement and make it available for use in California or Nevada. Arizona's decision to establish a water bank was clearly driven by the fact that Arizona is a junior user on the River and the will to 'secure' the unused portion of its 2.8 million acre-feet Colorado River allocation. The AWBA website clearly mentions:

> 'Until the AWBA was created, Arizona did not use its full 2.8 million acre feet share of Colorado River water. Without the AWBA, Arizona would not have used its full allocation until the year 2030. During that interim period, the accumulated amount of water left in the Colorado River would have amounted to approximately 14 million acre feet. Most of that water would have gone to southern California.
>
> Leaving a portion of Arizona's water in the Colorado River was a lost opportunity. The AWBA seizes this opportunity and gives Arizona the capability to further secure the dependable water supplies necessary to ensure the State's long-term prosperity'[4].

Thus, water is stored in various storage facilities, usually underground, even if no specific needs correspond to such water. Two main types of storage facilities have been developed: underground storage facilities (see picture below) and groundwater saving facilities. Underground storage facilities provide direct recharge of surface water via spreading basins, while groundwater savings programs creates storage credits through the use of surface water in lieu of pumping groundwater. Arizona has invested tens of millions of dollars in storage facilities.

In 2006, the AWBA had fifteen water storage permits for groundwater saving facilities, with a total storage capacity of 699,359 acre-feet, and ten water storage permits for undergrounds storage facilities, with a total storage capacity of 781,000 acre-feet per year (AWBA 2005). Thus, in total, AWBA can potentially store 1,480,359 acre-feet per year.

The picture below represents one of the largest underground storage facilities, the Agua Fria Recharge Project. It is located northwest of Phoenix in the flood plain of the Agua Fria River, about four miles below the location where the Central Arizona Project aqueduct crosses the river channel. It consists of four miles of in-stream channel recharge, a diversion facility and about 100 acres of infiltration basins. Its normal expected use is 20,000 to 60,000 a.f. per year.

[4] http://www.azwaterbank.gov/awba/Background_Summary.shtml, last visited in February 2010.

Figure 6 Agua Fria Underground Storage Facility (Credit to CAP).

Besides its primary objective to secure Arizona's unused Colorado apportionment, the creation of the AWBA was also intended from the beginning to allow interstate banking programs. Thus, the statutes creating the AWBA explicitly recognize the possibility for the AWBA to enter into 'interstate water banking agreements with appropriately authorized agencies in California and Nevada'. The statutes also provide the costs to be paid by the agencies in California or Nevada[5].

[5] 'The costs include all of the following:

1. The cost of acquiring Colorado River water.
2. The cost of delivering that Colorado River water through the central Arizona project to a storage facility, including fees for the operation, maintenance, pumping energy and capital costs of the central Arizona project as established by CAWCD.
3. Amounts equivalent to taxes ordinarily paid by CAWCD subcontractors and their customers to pay for the repayment, operation and maintenance costs of the central Arizona project, to the extent those equivalent amounts are not collected by paragraph 8 of this subsection.
4. The cost of storing that Colorado River water.
5. The cost of constructing, operating and maintaining a storage facility to the extent that facility stores water for the California or Nevada agency.

The price of water is not determined by the market but by Arizona's administration. Indeed AWBA purchases Central Arizona Project water at the administrative price set annually by the CAP and then adds operational, maintenance costs and other costs linked to the storage to the CAP price.

Finally, the statutes make reference to article II (b)(6) of the decree *Arizona v. California* and the role of the Secretary of the Interior to authorize the delivery of unused entitlement:

> 'In each water banking agreement, the authority shall agree that in years in which the California or Nevada agency requests recovery of water stored in Arizona, the authority shall cause a decrease in Arizona diversions from the Colorado River by the amount of water requested for recovery by the California or Nevada agency, thus creating unused entitlement for delivery to that agency by the United States Secretary of the Interior pursuant to article II(b)(6) of the decree'.

Federal Intervention: The Final Rule on Offstream Storage of Colorado River Water and Development and Release of Intentionally Created Unused Apportionment in the Lower Division States

By creating the AWBA, Arizona got the initiative over the establishment of an interstate banking system in the Colorado Basin. However, resting its legitimacy on the 1964 Decree, and recognizing the role of the Secretary of the Interior in the delivery of unused entitlements, the interstate banking program needed to be approved and governed by a federal rule. A first proposed rule was published by the Interior in December 1997 for public comments and a Final Rule titled 'Offstream Storage of Colorado River Water and Development and Release of Intentionally Created Unused Apportionment in the Lower Division States' was issued in 1999. This Rule establishes 'a procedural framework for the Secretary of the Interior to follow in considering, participating in, and administering storage and interstate release agreements in the Lower Division states'. 'The storage and interstate release agreements would permit state-authorized entities

---

6. The cost of recovering the stored water and delivering it to Colorado River water users in this state to use in place of Colorado River water that would otherwise be used.
7. A fee equivalent to the approximate amount of administrative, legal and technical expenses incurred by the authority in storing water for the California or Nevada agency, recovering that stored water and making an equivalent amount of Colorado River water available to the California or Nevada agency (Arizona Statutes 45-2471 C.)
8. Any fee paid in lieu of taxes pursuant to section 48-3715, subsection B by the authority in acquiring the water to be stored.'

to store Colorado River water offstream, develop intentionally created unused apportionment (ICUA), and make ICUA available to the Secretary for release for use in another Lower Division state' (Part 414, Federal Register, vol. 64 n°210, November 1,1999).

Intentionally Created Unused Apportionment-ICUA

The Final Rule thus created a new concept: 'Intentionally Created Unused Apportionment' – ICUA. Intentionally Created Unused Apportionment is unused apportionment within a lower division state's basic or surplus apportionment, or both, which is not put to beneficial consumptive use that year within that state (§414.2). For this water to qualify as Intentionally Created Unused Apportionment, the water must first be offered to all entitlement holders within the storing state for purposes other than interstate transactions under proposed storage and interstate release agreements (§414.3).

ICUA can be created either by the storing or consuming state for later release and use in the consuming state. In other words, if stored water belongs to the storing state, the interstate banking deal includes both storage and sale of water. If water stored belongs to the consuming state, the deal includes storage costs only.

The concept of ICUA founds its roots in article II B(6) of the 1964 consolidated decree in the case of *Arizona v. California* which makes the Secretary responsible for the release of unused apportionment :

> 'If, in any one year, water apportioned for consumptive use in a state will not be consumed in that state, whether for the reason that delivery contracts for the full amount of the state's apportionment are not in effect or that users cannot apply all of such water to beneficial uses, or for any other reason, nothing in this decree shall be construed as prohibiting the Secretary of the Interior from releasing such apportioned but unused water during such year for consumptive use in the other states. No rights to the recurrent use of such water shall accrue by reason of the use thereof' (Consolidated Decree 1964).

Authorized entities

Water transactions are limited to 'authorized entities'. The Final Rule is quite vague regarding the definition of 'authorized entities.' These entities are defined by state laws:

Authorized entity means:

(1) An entity in a storing state which is expressly authorized pursuant to the laws of that state to enter into storage and interstate release agreements and develop ICUA (storing entity); or

(2) An entity in a consuming state which has authority under the laws of that state to enter into storage and interstate release agreements and acquire the right to use ICUA (consuming entity). (§414.2, Final Rule, Part 414, Federal Register, vol. 64, n°210)

Thus, each state has the power to determine who should take part in water transactions. For the time being they are taking place at the state level and are limited to state authorities only.

When the Final Rule on interstate transaction was first released, Indian Tribes expressed their opposition on the ground that it does not provide specific and express protection of the Tribes' interests both in making water transfers and developing tribal water on or off their reservations. Indian Tribes commented that they should be permitted to market their unused water on the same basis as the states. The Department of the Interior answered that it is 'encouraging the Lower Division states to enact measures and take actions that will allow the tribes to participate in opportunities covered by this rule'. [. . .] But the Department of the Interior stressed further that 'no holders of Central Arizona Project allocations have a right to store this water for an interstate transaction unless they can qualify as an authorized entity under this rule.' The transactions emerging in the Lower Colorado Basin therefore involve primarily state authorities. Whereas one observes that market forces are likely to gain greater influence in the Lower Colorado Basin, questions such as whether the Tribes should be allowed to market water out-of-state are raised.

Financial Compensation

These new type of transactions clearly involve financial compensations. Indeed, the storing state is allowed to store the unused apportionment of the consuming state or its own unused apportionment for later use in the consuming state. Therefore, financial compensations include not only storage costs but also the price of water itself when the agreement comprises a sale of water. Interestingly the Secretary recognizes the existence of such financial transactions over Colorado River water, but clearly explains that it is Lower Basin states' matter.

The Final Rule explains:

> 'Financial details between and among the non-federal parties need not be included in the Storage and Interstate Release Agreement but instead can be the subject of separate agreements. The Secretary need not be a party to the separate agreements' 414.3(17).

The 1999 Final Rule on offstream storage and development of ICUA modified substantially the definition of consumptive use in the Law of the River. By allowing Lower Basin states (in fact, essentially Arizona) to store their unused apportionment for interstate purpose, the Secretary of the Interior recognized a greater degree of ownership to state's basic and surplus apportionments. If they have sufficient storing facilities, states can keep control over their entire Colorado water apportionment even though this water is not put to beneficial use within their state lines.

This evolution of the definition of consumptive use seems to contradict the provision of article I(K) of the 1964 decree, which stipulates that:

> 'Consumptive use of water diverted in one state for consumptive use in another state shall be treated as if diverted in the state for whose benefit it is consumed.'

In the case of Intentionally Created Unused Apportionment, the amount of stored unused water is counted as consumptive use in the storing state in the year it is stored and not in the consuming state. When ICUA are released, they are counted 'as consumptive use in the consuming state of unused apportionment under article II(B)(6) of the decree in the year the water is released in the same manner *as any other unused apportionment taken by that state*' (§414.4(b), emphasis added). In other words, released water by the storing state is not counted within the basic apportionment of the consuming state.

A few years after the adoption of the Final Rule on ICUA, Herb Gunter, the then-director of the Arizona State Department of Water Resources, commenting on the evolution of the Law of the River brought by the Final Rule, explained that it was 'the new way of dealing with water in the West'. 'We will continue to have some water in excess of what we should be able to put to beneficial use'. 'As long as we have that water and it can be put to beneficial use by another state in the basin, we have the right to store that water on their behalf in exchange for compensation'(Rake 2004). Such a comment corresponds to a new interpretation of the term 'beneficial use' suggesting that commercial use is now considered as 'beneficial use' (i.e. the sale of water to another state). As we will see, this evolution is far from being widely accepted.

Storage and Interstate Release Agreement Negotiations

It is important to look at the particular arrangements that were negotiated in the framework of the Federal Rule on offstream storage and the development of the ICUA. They indeed reveal the varied ways in which the Lower Basin states interpreted that Rule. Only two interstate banking agreements have been negotiated since 1999. They both are quite different in terms of size and form. The first agreement involves the storage of both Nevada and Arizona unused apportionments in Arizona for later use in Nevada. The second agreement provides that Nevada's unused apportionment be stored in Southern California by the Metropolitan Water District for future use in Nevada.

Interstate Banking Agreement between SNWA and AWBA

An Interstate Storage and Release Agreement was signed in 2002 among the Secretary of the Interior, the Colorado River Commission of Nevada (CRCN), the Southern Nevada Water Authority (SNWA), and the Arizona Water Banking Authority (AWBA). This agreement provides for the general rules and the role of the Secretary in releasing unused apportionment according to article II B(6). This agreement is accompanied by two other agreements. In the Interstate Water Banking Agreement, the SNWA and the AWBA agreed on their water transactions and financial compensations. Finally, the agreement between AWBA and the Central Arizona Water Conservation District (CAWCD) provides that CAWCD will use groundwater instead of Colorado River water, thus reducing its diversion from the Colorado and enabling the AWBA to store Intentionally Created Unused Apportionment.

The Interstate Banking Agreement between Arizona and Nevada allows the ArizonaWater Banking Authority (AWBA) to bank/store Arizona's and Nevada's unused Colorado River water to establish storage credits for the Southern Nevada Water Authority. Thus, depending on water availability, both Arizona and Nevada unused apportionments are stored in Arizona for future use by the Southern Nevada Water Authority (Maher 2006).

The 2001 water deal between Nevada and Arizona was later amended in 2004. When Nevada negotiated the original agreement, it anticipated that its water needs would be met through 2016 by surplus water made available through the Interim Surplus Guidelines (ISG)[6]. After 2016, Nevada intended to utilize its credits stored by AWBA. However, as the surplus water were allocated depending on the availability of flows and elevation of Lake Mead, and as in 2004 surpluses under the ISG were not available because of the enduring drought, Nevada had to find other ways to respond to its water needs. Therefore, Nevada approached Arizona with a proposal to amend the existing agreement. The 2004 amendment significantly modifies the 2001 agreement. In the original agreement, AWBA

[6] The ISG published on January 25, 2001 by the United States Department of the Interior, control the operation of the Lower Colorado River system through 2016. The ISG provide more liberal surpluses to allow a transition period for California to reduce its reliance on Colorado River water and to implement measures to limit its total uses of Colorado River water in the State to 4.4 million acre-feet. Consequently, at the time the 2001 Agreement was negotiated, Nevada anticipated that its water needs would be met through 2016 by surplus water made available through the ISG.

promised its 'best effort' to store sufficient supplies of Colorado River water to enable Nevada to pay for and earn 1.25 million acre-feet of long-term storage credits. The amendment changes 'best effort' to 'guarantee' of specific annual delivery amounts. Nevada receives 20,000 acre-feet of water in 2007 and 2008, 30,000 acre feet in 2009 and 2010 and 40,000 acre-feet in 2011 and beyond. Nevada has until 2060 to exhaust its Arizona water credits (i.e. 1.25 million acre feet).

In exchange for this guarantee, Nevada paid Arizona $100 million above the actual cost of water storage and delivery in 2005. Arizona then receives $230 million for actual cost of water delivery and storage by annual payments of $23 million, starting in 2009. The amendment reduces the annual recovery obligation. Under the 2001 agreement, Nevada could request recovery up to 100,000 acre feet in any given year (123 million m^3). The amended agreement's upper limit is only 40,000 acre feet (50 million m^3) except during shortage years[7].

By creating an interstate banking program with Nevada, Arizona wanted first to obtain additional funding to develop its storing facilities. It was also a way to secure its unused apportionment. AWBA's interstate accounting report of 2007 clearly mentioned:

> 'Without interstate banking, the AWBA does not have sufficient in-state funding to store the quantities of excess CAP water currently projected to be available. Interstate water banking is a mechanism that can maximize Arizona's utilization of its 2.8 million acre-feet entitlement' (AWBA 2007, 4).

Moreover, the interstate banking agreement between Nevada and Arizona encompassed an important political dimension. Arizona indeed got a commitment in writing from Nevada to support Arizona's call for change in the existing rules governing the Colorado River in case of a shortage. Arizona's prime concern was indeed the junior status of its Colorado water entitlements. Under current rules, in case of shortage, Arizona's allocations would be the first to be cut. The larger context of the creation of the Arizona interstate banking program and the negotiation of the interstate banking agreement is indeed the discussion over a drought plan for the Colorado Basin and the order of priority of water rights in case of scarcity. AWBA's interstate accounting report of 2007 makes it clear that:

> 'Most importantly, the Amended Agreement helps strengthen Arizona's interstate relationship with Nevada. A strong relationship is critical as the seven Basin states

[7] On average, Nevada will receive 50 million m^3 each year over 30 years (a total of 1.25 m.a.f.: 1,5 billion m^3). The volume of water transferred between Arizona and Nevada is comparable with the project between Turkey and Israel.

continue discussions regarding the drought and shortages on the Colorado River (AWBA 2007, 4).

We will later study the outcomes of the recent negotiations over the Colorado drought plan.

Interstate Banking Agreement between MWD and SNWA

AWBA reports that discussions took place between the AWBA and the Metropolitan Water District (MWD) of Southern California on storage and interstate release agreements, but that eventually no agreement was signed (as of 2009). The only arrangement that exists between the two entities results from a demonstration program dating back to 1992 between CAWCD and the MWD. A total of 80,909 a.f. is to be recovered by the MWD until 2010.

If there is no agreement under the Final Rule between MWD and AWBA, there is, however, an interstate banking agreement between the MWD of Southern California and SNWA negotiated in 2004. This later agreement has a much smaller dimension than the agreement between AWBA and SNWA.

MWD will store a portion of Nevada's unused Colorado River water in Southern California until it is needed. Nevada will be authorized to recover up to 30,000 acre-feet per year from the storage account, with six months advance notice provided to MWD. It is reported that SNWA banked 10,000 a.f. in 2004 and another 10,000 a.f. in the California Water Bank in 2005.

The 1999 Final Rule well suits Arizona's interests. Indeed, among the Lower Division states, Arizona is the only one that still has an important amount of unused apportionment. It is therefore more interesting for Arizona than for any other Lower Basin state to be allowed to store its unused water for future consumptive use in another state.

The development of offstream storage facilities is not new in the Colorado Basin, but their storing capacities have reached unprecedented levels. AWBA estimated that in 2007 the total amount of water stored in Arizona was equivalent to Arizona's full annual apportionment of the Colorado River (approximately 2.8 m.a.f.). Storage facilities are becoming more important to water managers every day in the Lower Basin states, and various private companies are investing in this field. For example, Cadiz Inc. has constructed an underground storage facility of one m.a.f. capacity in Southern California (http://www.cadizinc.com/). SRP (Salt River Project) is one of the biggest storing companies in Arizona, operating the Granite Reef Underground Storage Project (200,000 a.f.), and is currently developing another storage facility called the New River - Agua Fria River Underground Storage Project (75,000 a.f.) (http://www.srpnet.com).

With the rapid development of underground storage projects, new problems arose. A judicial case in Arizona addressed the question of private land used for banking purposes (*West Maricopa Combine, Inc. v. ADWR*, 1 CA-CV 00-0086, June 5, 2001). The Arizona Court of Appeal held that Arizona water statutes (A.R.S. § 45–173) explicitly authorize the use of natural channel to convey water, and therefore an applicant for an underground storage facility permit does not need the consent of a landowner whose property is traversed by the natural channel being used for the storage facility (Office of the Special Master 2001).

The Court held:

> 'Based on Arizona's enduring and fundamental policy encouraging the full use of scarce water resources and the plain language of A.R.S. § 45–173, we hold that real property owners such as 10K may not prevent a beneficial user from using an existing natural watercourse for water storage purposes'(*West Maricopa Combine, Inc. v. ADWR*, 1 CA-CV 00-0086, June 5, 2001).

The concept of 'underground water storage' - in artificial reservoirs or through artificial recharge of natural but depleted aquifers – is a specific phenomenon to the Western United States - at least in its proportion. Storage and interstate release agreements have participated in the developments of such facilities. The terms of 'water storage' and 'water banking' sound highly inappropriate to most conceptions of sustainable management of watercourses and their ecosystems (at least in US proportions). Besides, they reveal how Colorado River water apportionment according to the Law of the River is perceived by most basin states, that is as a quantity of water that each state owns whether they need it or not. To understand how the Colorado River water apportionment is perceived, it is important to study the negotiations over the drought plan that was adopted in 2007. In fact, it seems that the concept of ICUA and the resulting evolution of the definition of consumptive use did not really convince all basin states. In many respects, the drought plan took another direction in the management of the Colorado River especially by introducing a very different type of mechanisms: the Intentionally Created Surplus.

NEGOTIATIONS OVER THE DROUGHT PLAN: 'OPERATION OF LAKE MEAD AND LAKE POWELL UNDER LOW RESERVOIR CONDITIONS'

With recurring drought conditions and the implementation of the Interim Surplus Guidelines, the State of Colorado along with the other Upper Basin states required a revision of the operation of Lakes Mead and Powell and the definition of special rules under 'low reservoir conditions' in 2005.

Before analysing the negotiations over the drought plan which was adopted in 2007, it is necessary to quickly recall the context that led to the adoption of Interim Surplus Guidelines (ISG).

California's Quantification Settlement Agreement and Interim Surplus Guidelines 1996–2003

From about 1990 to 2002, the central problem was how to bring California's use of Colorado River water from 5.2 million acre-feet per year (m.a.f.) down to its basic, normal year apportionment of 4.4 m.a.f.

In 2001, in order to facilitate the negotiations on a Quantification Settlement Agreement (QSA), a framework - known as the Interim Surplus Guidelines - was adopted by the Interior, California and the other six basin states. The Interim Surplus Guidelines offer California a 'grace period' of 15 years to gradually reduce its use of the Colorado River if a QSA was reached among the Californian agencies before December 2002, and if the QSA met all the guidelines' benchmarks. However, if California failed to meet the deadline, it would immediately lose access to that grace period of extra water. The Metropolitan Water District of Southern California, the San Diego County Water Authority, the Coachella Valley Water District, and the Imperial Irrigation District (IID) finally agreed on a plan in October 2003 whereas the first deadline of December 2002 was already behind. After having cut for one year the surplus water to California, the US Department of the Interior decided to reinstate the provisions of the Interim Surplus Guidelines.

California 4.4 Plan

The plan relied heavily on water conservation in the agricultural sector and water transfers to the urban sector. Under the plan, up to 800,000 a.f. of water would be conserved via dry-year fallowing agreements, canal seepage recovery, groundwater banking, conjunctive use and desalination of drainage water, and an American Indian water rights dispute settled within the state.

The biggest transfer of the QSA is the one between IID and San Diego of up to 200,000 a.f. annually. Imperial Valley farmers served by IID signed up in late 2003 to fallow land or conserve irrigation water to free up water for the transfer.

The key issue in the QSA negotiations was the fate of the Salton Sea, which is sustained by irrigation runoff from the Imperial Valley and Coachella Valley farms. It was decided that IID sell surplus Colorado water to the state for 175$ per a.f., which in turn would sell it to MWD for 250$ per a.f. The profit realized on the sale was used to fund Salton Sea restoration work.

The Upper Basin States Asked to Reduce their Deliveries to Lower Basin States

In the years following the adoption of the ISG, drought conditions in the Colorado River Basin caused a significant reduction in storage levels in Lakes Powell and Mead, and precipitated discussions among and between the basin states and the Department of the Interior over the adoption of shortage guidelines.

In 2005, the Upper Basin states, led by the State of Colorado, required a revision of the Annual Operating Plan (AOP) of Lakes Mead and Powell. In a letter to the Secretary of the Interior, they expressed their concerns regarding the drop in Lake Powell's level and asked for a reduction of the volume of water delivered by the Upper Basin states to the Lower Basin states. The volume of water released from Lake Powell has been established at 8.23 m.a.f. for more than three decades (1970 Long Range Operating Criteria)[8]. The Upper Basin states argued that such a quantity did not correspond to an obligation under the Law of the River. They denounced Arizona's strategy of storing water while the Upper Basin was experiencing a severe drought.

They wrote:

> 'Given the abundance of water in the Lower Basin as compared to the drought conditions that continue to persist in parts of the Upper Basin, as well as the relative difference in storage in Lakes Powell and Mead, it would be prudent water management to hold as much water in Upper Basin reservoirs as possible' (Letter to Norton, April 18, 2005).

For Upper Basin states, the 1922 Compact provision requiring that they deliver an average of 75 m.a.f. every ten years offers them some flexibility for the delivery of water to the Lower Basin states. According to them, this provision allows them to release less water in some years of severe drought and more water when available.

Behind these concerns lies the question of the protection of the full apportionment of the Upper Basin states. For the time being, Upper Basin states have not yet developed their full apportionment under the Colorado River Compact and are looking for a way to secure their entitlements.

This profound disagreement between Upper and Lower Basin states led the Secretary of the Interior to ask the Bureau of Reclamation to develop interim

[8] 8.23 m.a.f. corresponds to the apportionment to the Lower Basin states (7.5 m.a.f.) plus half the quantity of water delivered to Mexico (1.5 m.a.f./2). The Compact provides indeed that the 1.5 m.a.f. delivered to Mexico be equally shared by Upper and Lower Basin states (article III (b) of the 1922 Compact). To this sum (8.25 m.a.f.), 20 000 acre feet are retained because the Paria River, a Upper Basin River, flows into the Colorado River below Glen Canyon Dam (Lake Powell) (8.25-0.2=8.23 m.a.f.).

guidelines to address Lower Basin shortage and operations of Lake Powell and Lake Mead under low reservoir conditions (for an interim period until 2026). It is considered that there is a shortage when there is insufficient mainstream water available to satisfy 7.5 m.a.f. of annual consumptive use in the Lower Division states (article II B (3) of the 1964 consolidated decree).

To fulfil its tasks, the Bureau of Reclamation initiated an Environmental Impact Statement (EIS) supposed to evaluate all the consequences of different shortage guidelines proposals on water storage, power production, navigation, recreation, socio-economic conditions and the environment. Public participation was solicited by the Bureau of Reclamation during the EIS process, which held a series of public meetings in the different basin states. Proposals and comments were made by a wide range of interested parties that included federal, state and local agencies, as well as Indian tribes, businesses, NGOs and individuals. The different documents pertaining to the EIS process were regularly posted on the Bureau of Reclamation's webpage.

In the EIS process, five alternatives were reviewed: the no action alternative, the seven basin states' proposal, the conservation before shortage proposal, the demand supply alternative and the reservoir storage alternative. Among these alternatives, three emanated from the Bureau of Reclamation, one came from the seven basin states and one came from a consortium of NGOs (the 'conservation before shortage' proposal). Among these alternatives, the seven basin states' and the NGOs' proposals received the most attention, and the preferred alternative of the final EIS combined elements of both proposals. We now turn to an analysis of the major characteristics of the preferred alternative published by the Bureau of Reclamation in November 2007. A Record of Decision, endorsing the preferred alternative, was signed by the Secretary of the Interior Dirk Kempthorne on December 13, 2007.

Preferred Alternative

Although the Law of the River provides for some guidance in case of shortage in the Lower Basin, there remain many unclear aspects, especially concerning the coordination between Upper and Lower Basin states and responsibilities relating to water deliveries to Mexico. The adoption of shortage guidelines was thus necessary, but these new rules unfortunately remain deeply embedded into the Law of the River. The centuries-old rule of prior appropriation is not questioned. There is still no creation of a permanent decision-making body for the entire basin composed of representatives of each state and other interest groups of the river basin. During the negotiation process, the states seemed to negotiate over ownership rights to certain water quantities and not over authorized uses and the sustainable management of a commonly-owned resource. This is due to the lack of definition of the common property regime of the Colorado River water. The property regime of the Colorado River is well illustrated by the status of apportioned but unused water, considered to be private ownership of states. As we saw earlier, Arizona is allowed to store/'bank' the unused part of its apportionment. Under these circumstances, allocation for the environment will always be neglected.

The 2007 Record of Decision, however, opens some new avenues for the management of the Colorado River Basin. The concept of Intentionally Created Surplus (ICS) is, in this regard, particularly noteworthy. Before analysing the significance of such a mechanism, we will present the other important provisions of the new guidelines.

Allocation of reduced water deliveries

The 2007 Guidelines describe how the reduction in water deliveries from Lake Mead to Lower Basin states will be allocated. These reductions will depend on the elevation of Lake Mead (reduction of 333, 000; 417, 000 and 500, 000 a.f. from Lake Mead at elevations 1,075, 1,050, and 1,025 feet, respectively[9]). The Guidelines also provide how to coordinate operations of Lake Powell and Lake Mead through the full range of reservoir levels. This means, concretely, that the Upper Basin states will not have to release 8.23 m.a.f. from Lake Powell each year, but will be allowed to release more or less than this volume depending on the level of both reservoirs.

Shortage sharing agreement between Arizona and Nevada

The 2007 deal encompasses a shortage-sharing agreement between Arizona and Nevada. Article II B(3) of the consolidated decree of 1964 *Arizona v. California* clearly recalls the prior use doctrine in the Lower Colorado River Basin:

> 'If insufficient mainstream water is available for release, as determined by the Secretary of the Interior, to satisfy annual consumptive use of 7,500,000 acre feet in the aforesaid three states, then the Secretary of the Interior, after providing for satisfaction of present perfected rights in *the order of their priority dates* without regard to state lines [. . .] may apportion the amount remaining available for consumptive use' (emphasis added).

[9] If Lake Mead falls below elevation 1,075 feet, the Secretary will consult with the basin states to determine how to allocate additional reductions.

Arizona and Nevada both have junior rights in comparison with California; therefore the reduction of water deliveries from Lake Mead will be shared by Arizona and Nevada only. California's entitlement will not be reduced in case of shortage (at least up to 500,000 a.f. of reduction). Thus, Arizona and Nevada decided to agree on a way to share the reduction among them (if Lake Mead elevation is below 1,075 feet, Arizona's entitlement is reduced by 320,000 a.f. and Nevada by 13,000 a.f.; below 1,050 feet: Arizona 400,000 a.f., Nevada 17,000 a.f.; below 1,025 feet: Arizona 480,000 a.f., Nevada 20,000 a.f.).

The shortage sharing agreement also mentions that the Southern Nevada Water Authority agrees to provide the sum of eight million dollars to the Arizona Water Banking Authority (AWBA) 'for the purpose of assisting Arizona in offsetting impacts from shortages that may occur during the Interim Period'. The agreement explains that the AWBA will use this money to purchase and/or store water supplies' (article 1.5). This shortage sharing agreement is the illustration of the special relationship linking Arizona and Nevada following their interstate banking arrangement.

It is important to mention that this shortage sharing agreement between Arizona and Nevada - which is a key part of the seven basin states' proposal-, was conditioned in the states' proposal to the reduction of US water deliveries to Mexico (article 2).

Reduced Deliveries to Mexico in the Basin States' Proposal

In their Letter of April 2007 to Secretary Kempthorne, the seven basin states wrote:

> 'However, in the course of these negotiations no issue has surpassed the importance of how the United States exercises its authority to reduce the quantity of water allotted to Mexico under article 10(b) of the Mexican Treaty of 1944. [. . .] The basin states strongly urge the United States to exercise its authority to reduce the quantity of water allotted to Mexico in years in which the Secretary imposes shortages in deliveries of water from Lake Mead in the United States. . . ' (Letter to Kempthorne, April 30, 2007).

Article 10 (b) of the 1944 Treaty between the USA and Mexico concerning the utilization of the Rio Grande, the Colorado and Tijuana Rivers mentions that:

> '. . . In the event of extraordinary drought or serious accident to the irrigation system in the United States, thereby making it difficult for the United States to deliver the guaranteed quantity of 1,500,000 a.f. a year, the water allotted to Mexico under subparagraph (a) of this article *will be reduced in the same proportion as consumptive use in the United States* are reduced' (emphasis added).

The basin states proposed to calculate the percentage of reduction to Mexico according to the proportion of river flow Mexico gets in a normal year (this is different from the rule of article 10 (b)). They explained: 'the total quantity of water that will not be released or delivered to Mexico shall be based on Lower Basin water deliveries during normal water supply conditions. The proportion of the shortage that shall be borne by Mexico will be 17% (1.5 m.a.f. / 9 m.a.f. x 100% =17%)' (Letter to Norton, February 2006, Seven Basin States' Preliminary Proposal Regarding Colorado River Interim Operation).

Such a method to calculate the percentage of reduction to Mexico does not correspond to the provisions of the 1944 Treaty. Indeed article 10 (b) of the 1944 water Treaty provides that the deliveries to Mexico be *reduced in the same proportion as consumptive use in the United States is reduced*. Therefore the following calculus seems more appropriate than the basin states' proposal:

In the case of a shortage corresponding to an elevation at Lake Mead below 1,025 feet, the Lower Basin states' consumptive use is supposed to be reduced by 500,000 a.f. (reduction of 480,000 a.f. in Arizona and 20,000 a.f in Nevada, 0 in California). This reduction corresponds to a reduction of 6.7% of the normal deliveries to the Lower Basin states (500,000x100/7,500,000=6.7).

Therefore, in the event article 10(b) is applied because the conditions are considered to be equivalent to an 'extraordinary drought', the reduction of water deliveries to Mexico could equal 6.7% of the normal year 1,500,000 a.f. (and not 17% as proposed by the basin states[10]). However such a calculus is still wrong with regards to article 10(b) of the 1944 Treaty.

The Mexican section of the International Boundary Water Commission reacted strongly to draft EIS released in February 2007, and sent their comments to the Bureau of Reclamation in April 2007. They recalled:

> 'Any reduction in the allocation of water to Mexico shall be done in strict conformance with the terms of the Treaty. As stated in the meetings, Mexico views that the reduction applicable to Mexico in the event of extraordinary drought should be proportional to consumptive uses in all of the basin states, *not just those of the Lower Basin*' (Letter of the Mexican section of IBWC dated April 30, 2007, EIS 2007, volume IV, F5).

The point at issue here is how the Upper Basin states participate in the reduction of water deliveries. Under the Law of the River, the guaranteed 1,500,000 a.f. to Mexico by the US is divided into two equal contributions of 750,000 a.f. by both

---

[10] The difference between the two methods corresponds to a difference of 154,500 a.f. (basin states' proposal: 17% of 1,500,000 = 255, 000 a.f.; other method corresponding to article 10(b): 6.7% of 1,500,00 = 100, 500 a.f.).

the Upper Basin and the Lower Basin. Therefore, the former calculus should be modified to determine the exact proportion of consumptive use's reduction in each of the two basins and then aggregate them to find the true proportion of water use reduction in the United States (and not just in the Lower Basin). If we estimate that the Upper Basin states do not reduce their consumptive use and continue to contribute to the international commitment of the US by releasing 7,500,000 a.f. to Mexico, the proportion of deliveries reduction to Mexico should correspond to 6.7% of 750,000 a.f. and not 6.7% of 1,500,000 a.f. According to this calculus, if an 'extraordinary drought' were declared in the Colorado River Basin, the deliveries to Mexico would be reduced by 50,250 a.f.; this corresponds to the exact proportion of consumptive use reduction in the US (0 reduction in the Upper Basin; 6.7% in the Lower Basin).

The Mexican section of the IBWC and the *Comision Nacional del Agua* (CNA) underlined that if a reduction of water deliveries to Mexico were to be calculated this should be done only when reduction were applied to the entire US Basin not just in Arizona and Nevada.

> 'In the modelling of the reductions, Mexico is always included with Arizona and Nevada, while California is not included until level 2 is reached, and the Upper Basin is never included. This generates false expectations as to the timing and conditions under which there would be reductions to Mexico as well as the implicit acceptance by Mexico of those reductions' (letter of the Mexican section of IBWC).

Besides, the CNA mentioned that before the reductions to Mexico be calculated, it will be first necessary to define the term of 'extraordinary drought'. Noticing the reduction of Colorado runoff in the US but no significant decrease of precipitation, the CNA suggested that the relationship between groundwater withdrawals in the US and river flow be better studied.

In its final EIS, the Bureau of Reclamation did not take a clear position neither on the controversial issues of the impact of the shortage guidelines on Mexico nor on the potential reduction of water deliveries to Mexico. The final EIS is very prudent on that matter:

> 'In order to assess the potential effects of the proposed federal action in this Final EIS, certain modelling assumptions are used that display projected water deliveries to Mexico. Reclamation's modelling assumptions are not intended to constitute an interpretation of application of the 1944 Treaty or to represent current United States policy or a determination of future United States policy regarding deliveries to Mexico.
>
> The United States will conduct all necessary and appropriate discussions regarding the proposed federal action and implementation of the 1944 Treaty with Mexico

> through the IBWC in consultation with the Department of State' (Bureau of Reclamation 2007, 18).

Further the preferred alternative of the EIS lets open the possibility for Mexican entities to participate in the ICS program.

Intentionally Created Surplus Program

The most innovative aspect of the 2007 Guidelines is the introduction of a new concept: 'Intentionally Created Surplus' (ICS). The purpose of these ICS is to augment storage of water in Lake Mead and the entire system in general. ICS are supposed to increase the flexibility of Colorado River management. Different entities will be allowed to store/bank ICS credits in Lake Mead for later use. These ICS credits will be created through extraordinary conservation, system efficiency, tributary conservation or importation of non-Colorado River water. ICS credits can only be created if such water would have otherwise been beneficially used (contrary to intentionally created unused apportionment, ICUA). No category of surplus water can be used to create ICS credits (contrary to ICUA). ICS credits will be subject to a tax: 5% of the ICS credits will be dedicated to the system (one time) and another tax of 3% due each year to compensate evaporation resulting from storage in Lake Mead. The basin state's proposal provided that ICS credits only be released to the entity that created it, whereas the Conservation before the shortage's proposal allowed for ICS trading among willing entities. The 2007 Guidelines contain only vague principles regarding the exact conditions for the release of these ICS credits.

The concept of ICS is justified under the Law of the River by article II(B)2 of the consolidated decree, which applies to water 'in excess of 7,500,000 a.f.' Article II(B)2 mentions:

> 'If sufficient mainstream water is available for release, as determined by the Secretary of the Interior, to satisfy annual consumptive use in the aforesaid states in excess of 7,500,000 acre feet, such excess consumptive use is surplus, and 50% thereof shall be apportioned for use in Arizona and 50% for use in California; provided, however, that if the United States so contracts with Nevada, then 46% if such surplus shall be apportioned for use in Arizona and 4% for use in Nevada'.

For the ICS to remain available for the entity that created them and not released for use by another entity in another Lower Basin states according to the percentages in article II(B)2, their creation should be accompanied by 'forbearance agreements'.

In forbearance agreements, Colorado water contract holders forgo (or refuse to exercise) their rights to surplus water under article II(B)2.

The interim guidelines provide for different categories of ICS. In fact, tributary conservation ICS, imported ICS and system efficiency ICS can be converted into 'extraordinary conservation ICS'.

'Extraordinary conservation ICS' can be created through activities such as land fallowing, canal lining and desalination (the desalinated water is used in lieu of Colorado water). But they can also be created by conservation on Colorado tributaries (if the water rights date to prior to 1929). ICS credits can also be created by financially contributing to federal projects designed to save water 'that would otherwise be lost' ('system efficiency ICS'). Finally, ICS credits can be created by importing non-Colorado River system water into the Colorado River mainstream ('imported ICS').

The first ICS programs and projects

In 2006 and 2007 the Metropolitan Water District of Southern California (MWD) and the Imperial Irrigation District (IID) both had ICS demonstration programs with the Bureau of Reclamation. These demonstration programs were undertaken in order to study whether ICS could be used as a long-term tool for water management in the Lower Basin.

Under a 2006 agreement with the Bureau of Reclamation, the MWD was allowed to store up to 200,000 a.f. in 2006 and 2007 in Lake Mead. These ICS credits were created through extraordinary conservation measures, essentially land fallowing implemented by the Palo Verde Irrigation District (PVID). In other words, the MWD had entered into fallowing agreements with owners of land within PVID and, instead of using this conserved water, the MWD decided to store it in Lake Mead as ICS water. This ICS water was subject to a system tax (5%) and an evaporation tax (2.8%).

The IID entered into a similar demonstration program in 2006. The Imperial Irrigation District created 5,000 a.f. of ICS credits in 2006 and 25,000 a.f. of ICS credits in 2007 through land fallowing programs. These credits were subject to the same taxes as MWD credits.

Even though the results of these two ICS demonstration programs were not known, the Bureau of Reclamation entered into other ICS agreements in 2007 and 2008 following the adoption of the 2007 Record of Decision.

In exchange for the funding of the construction of an 8,000 a.f. storage reservoir near the All-American Canal, the Southern Nevada Water Authority (SNWA), the MWD and the Central Arizona Water Conservation District (CAWCD) will receive in total 600, 000 a.f. of 'System Efficiency ICS credits' available through

2036[11]. It is the Bureau of Reclamation which will design and build the facility. On its webpage, the Bureau of Reclamation presents the project in the following terms:

> 'It is a system efficiency project that will help conserve Colorado River water that would otherwise not be put to beneficial use in the United States. Conserving this water and making it available to users in the United States will reduce the amount of water that must be released from storage at Lake Mead'[12].

This means, in other words, that the purpose of this reservoir is to reduce the volume of water delivered to Mexico at Morelos Dam in excess of the U.S.'s obligations under the 1944 Treaty. Such reductions will probably be very damaging for the Colorado Delta, which is already under strong pressure. The project raises other concerns. In contrast with other types of ICS credits, these credits will be created and released even before water is actually saved. According to the agreement, MWD was the first to use its ICS credits, in 2008, even though the reservoir was not yet constructed and the benefits for the efficiency of the system were not yet guaranteed. The project started in fall 2008 and was to be completed in 2010. It was planned, however, that the MWD take delivery of 34,000 a.f. of system efficiency ICS in 2008 and another 34,000 a.f. in 2009 (Bureau of Reclamation 2008, p25).

The Southern Nevada Water Authority is currently developing two other ICS projects. SNWA plans to create 'imported ICS credits' by conveying 9,000 a.f. per year of Coyote Spring Groundwater Rights to Lake Mead, where they would be banked for future use. The pipeline and facilities to transfer this water to the Muddy River and Lake Mead are scheduled to be completed by 2010 (SNWA 2008, p 25).

Another important project concerns the creation of 'tributary conservation ICS' from the Muddy and Virgin Rivers. As both rivers contribute to the Colorado River flow, only water rights that have a priority date preceding 1929 - the effective date of the 1928 Boulder Canyon Project Act - can be used for the creation of ICS tributary credits. In 2009, the Bureau of Reclamation approved the creation of 30,000 a.f. of tributary conservation ICS by Nevada. (SNWA 2009, p23).

---

[11] In a 2007 agreement with the Bureau of Reclamation, SNWA was allowed to receive 600, 000 a.f. of system efficiency credits in exchange for $172 million corresponding to the funding of the construction of Drop 2 storage reservoir (maximum annual delivery of 40,000 a.f. until 2036, beginning in 2011). Later, in May 2008 the MWD and the CAWCD joined the agreement. They will each receive 100, 000 a.f. in exchange for a financial contribution of $28.7 million (1/6 of the total funding) (therefore, SNWA's credits are reduced to 400,000 a.f.).

[12] http://www.usbr.gov/lc/region/programs/drop2reservoir.html, last visited in February 2010.

Toward ICS trading in the Colorado River Basin?

Most of the projects we just described have some market-based components. The MWD agreement with the PVID and the SNWA's transfers of water rights from the Coyote Spring Valley, Muddy Rivers and Virgin Rivers are cases where the entity that saves or conserves the water is not the entity that creates the ICS credits. This type of agreement is allowed in the Record of Decision adopted in December 2007:

> 'Extraordinary Conservation ICS from a project within a state may only be credited to the ICS Account of a Contractor within that state that has funded or implemented the project creating ICS, or to the ICS Account of a Contractor within the same state as the funding entity and project and with written agreement of the funding entity' (2007 ROD section XI.3.B. 8).

For the time being, this form of ICS water trading and banking is limited to contractors located in the same state. In the years to come, this type of ICS program will probably involve interstate ICS selling and buying with Lake Mead operated as a truly interstate ICS Bank. In addition, it should be noted that, in general, market-based options for water management in the Lower Basin are not new, and are increasingly being further developed in the region.

Thus, the Bureau of Reclamation recently launched the Water 2025 Challenge Grant program, whose objective is to contribute to 'projects that will stretch water resources through conservation, efficiency and water marketing'. Water marketing is a key tool of Water 2025. The 2005 Water 2025 report clearly mentions:

> 'The value of market-based approaches as an alternative is proven by the success of CalFed, the new Klamath water bank, the operation of the Central Valley Project in California, the ag-to-urban transfers in Southern California, and the 50 year-old water market in Northern Colorado' (Bureau of Reclamation 2005, 4).

The SNWA is also clearly open to market-based solutions in the Colorado River Basin. SNWA proposes the development of 'exchanges and transfers between willing sellers and willing buyers' in the Colorado River Basin. The 2008 Nevada Water Resource Plan mentions, 'for example, Southern Nevada could pay California to construct and operate desalination facilities in exchange for an equivalent portion of California's Colorado River water at Lake Mead'. Further, 'there also is the opportunity for interstate transfer of tribal water, but this issue still needs considerable discussion and agreement' (SNWA 2008, p29). SNWA also explains:

> 'While Southern Nevada does not have large volumes of Colorado River water available to transfer from agricultural to municipal use, there is the potential for

future agreements with other Colorado River Basin states. Similar to Nevada's agreement to bank water in Arizona future agreements could allow for agricultural transfers with other states, potentially increasing the availability of additional Colorado River water to Nevada'.

In a Senate hearing on water issues, Timothy Brick, the Chairman of the Board of the Metropolitan Water District of Southern California (MWD), recently declared that he was ready to 'bank' and 'trade' ICS in Lake Mead (California Institute for Federal Policy Research 2007).

Given this favourable context to market-based mechanisms, it is highly probable that an ICS trading scheme will sooner or later be introduced in the Lower Colorado River Basin. If Mexico participates in such a program, it will be the first-ever international water market to be created along an international watercourse.

Several forms of interstate ICS trading schemes could be created depending, among other things, on the participants and the size of the scheme.

For the time being, the 2007 Record of Decision provides that the total amount of ICS credits that can be accumulated in Lake Mead is 2.1 m.a.f. The following table summarizes the maximum amount of ICS credits that can be created per year and per state, the maximum amount of ICS credits that can be accumulated at any time for each state and the maximum amount of ICS credits that can be delivered per year and per state (2007 ROD section XI.3).

	Maximum annual ICS creation	*Maximum cumulative ICS*	*Maximum annual ICS delivery*
California	400, 000 a.f.	1.5 m.a.f.	400, 000 a.f.
Nevada	125, 000 a.f.	300, 000 a.f.	300, 000 a.f.
Arizona	100,000 a.f.	300, 000 a.f.	300, 000 a.f.
Total	**625, 000 a.f.**	**2.1 m.a.f.**	**1 m.a.f.**

Table 2 Volume limitations of ICS per state according to the 2007 ROD.

The final EIS allows for different scenarios concerning the quantities of ICS to be created and the entities allowed to participate in the ICS program. The final EIS thus provides for a maximum cumulative amount of 2.1 m.a.f. of ICS, but mentions this amount could be increased to 4.2 m.a.f., leaving open the possibility for 'other entities'(i.e. other than Colorado Water Contract Holders as defined in the section 5 of the 1929 BCPA) to participate in the ICS program (Bureau of Reclamation 2007, 20). The preferred alternative thus allows a doubling of the amount of ICS creation, accumulation and deliveries (see table below).

	Maximum annual ICS creation	*Maximum cumulative ICS*	*Maximum annual ICS delivery*
Entities (2007 ROD)	625, 000 a.f.	2.1 m.a.f.	1 m.a.f.
'Other entities'	625, 000 a.f.	2.1 m.a.f.	1 m.a.f.
Total	**1.2 m.a.f.**	**4.2 m.a.f.**	2 m.a.f.

Table 3 Volume Limitations of ICS in the Preferred Alternative.

In analyzing the possibility of extending the ICS program to other entities, the final EIS recognizes conservation before shortage's proposal as a potential option in the future. Indeed, the 'Conservation before shortage's alternative'[13] proposed the extension of the ICS concept to non-Colorado contract holders such as US federal agencies, state agencies, and private entities, including US NGOs, Mexican federal agencies and Mexican water users. Explaining that voluntary and compensated conservation is preferable to involuntary shortage, they proposed the introduction of market-based mechanisms into the ICS program. For example, the federal government, through the Bureau of Reclamation, should be allowed to purchase ICS credits for environmental purposes and other public benefit uses.

In addition, they recommended that negotiations over the extension of the ICS program be undertaken with Mexico. They wrote, 'we strongly suggest that the current federal process should leave the door open to the eventual approval of a bi-national extension of the ICS program, to limit the costs of future review of such a program and encourage the initiation of bi-national discussions about such a program' (Conservation before Shortage II, July 7, 2006). They further clarified that: 'it is critical to note that each of these programs could be undertaken without altering Mexico's basic Treaty entitlement to Colorado River water; any decrease (or increase) in Treaty deliveries would be matched by a mutually agreed-to compensation program or a proportionate increase (or decrease) in water deliveries in a later year. None of these programs would change Mexico's right to Colorado River water deliveries under the 1944 Treaty' ('Taking ICS to Mexico', Appendix K, final EIS 2007, p 32). They explained that the International Boundary Water Commission could be responsible for the implementation and monitoring of bi-national ICS programs. If Mexico participates in such a program, it could be the first-ever international water market to be created along an international watercourse.

[13] The Conservation before shortage alternative was proposed by a consortium of NGOs composed of Defenders of Wildlife, Environmental Defense, National Wildlife Federation, The Nature Conservancy, Pacific Institute, Sierra Club, Sonoran Institute, and Rivers Foundation of the Americas.

CONCLUSION

The experience of the Colorado River Basin is interesting because it highlights the difficulties in introducing flexibility into the management of an interstate watercourse. The Law of the River, based essentially on the 1922 Compact, is particularly rigid because the apportionments were made 'in perpetuity' between the two sub-basins. The average annual flow of the Colorado River was then divided into two equal parts of 7.5 m.a.f. At that time, and still today, no river basin commission was created for the entire basin. Apportionments were made in fixed quantities and not in percentages of river flow. The 1922 Compact made no explicit provision regarding the property regime of the Colorado River. As a consequence, the basin states considered that their apportionments to the river flow were their entire property. In 1928, by passing the Boulder Canyon Project Act, the US Congress made a 'permanent' apportionment of Colorado water among the Lower Division states. Congress also put the Secretary of the Interior in charge of managing and operating the Colorado River in the Lower Basin, but its role in fact remained very limited. Indeed, as the apportionments among the Lower Basin states were again made in fixed quantities of water and as the whole 7.5 m.a.f. were allocated to the states, the role of the Secretary in the management of the river was limited. At that time, the estimated annual average flow of Colorado River was entirely allocated for consumptive use in the basin states. No apportionment was made for the river system. In fact, it is up to each state to develop and implement environmental protection in accordance with federal rules. The Law of the River is also particularly inflexible because of the prior appropriation system and the 'use it or lose it' rule. Generally, the basin states are very attached to their apportionments under the Law of the River and, as we have seen, these apportionments were determined in a year where the river flow was high.

In order to protect its unused but apportioned water, Arizona decided to create huge storage facilities and establish the Arizona Water Banking Authority. In order to get more funding to develop its storage facilities and gain an ally in negotiations over Colorado River management in case of shortage, Arizona proposed to store its unused apportionment for later use in Nevada. An important evolution in the Law of the River happened once interstate banking was introduced, thus substantially modifying the definition of 'beneficial use'. With the issuance of the Federal Rule on 'offstream storage and development of ICUA', a state that stores its unused apportionment for later use in another state is recognized as putting its apportionment to beneficial use. In other words, commercial use (i.e. the sale of water to another state) becomes beneficial use.

With the adoption of this rule, the nature of Colorado entitlement evolved to an even greater degree of state ownership.

It seems that the adoption of the rule on ICUA did not satisfy all of the basin states. And, in fact, it essentially concerned Arizona. Indeed, Arizona is the only Lower Basin state that still has an important amount of apportioned but unused water. The adoption of the rule on offstream storage and development of ICUA offered few opportunities to the other basin states. Moreover, some states, essentially the Upper Basin states, did not adhere to the logic behind the interstate banking scheme. They had preferred a scheme more oriented toward water conservation and water efficiency. In many respects, Arizona's huge storing facilities go against the broad understanding of sustainable water management. In any case, with the decreasing Colorado River flow and the diminishing levels in Lake Mead and Lake Powell, the concept of ICUA became even more inadequate.

The new concept of ICS that emerged during the negotiations over the interim shortage guidelines bears more prospects for a true improvement of the management of the Colorado River. The creation of ICS clearly offers possibilities for increased water conservation and efficiency. ICS essentially concerns water conserved through 'extraordinary measures,' water that would have otherwise been used. The concept of ICS is in sharp contrast with that of ICUA. ICS credits will be stored in Lake Mead and will be governed by the federal government, whereas IUCA are managed by the states (essentially Arizona). Contrary to ICUA, ICS will not be applicable to surplus water, but to a state's basic apportionment only. Finally, contrary to ICUA, ICS credits are subject to a tax to benefit the whole system, thus increasing the role of the Bureau of Reclamation in monitoring the scheme and managing the river flow.

The ICS program will certainly provide for more flexibility in the management of the Colorado River, especially if non-Colorado contact holders are allowed to participate. As we saw earlier, it is highly probable that in the near future ICS credits will become tradable. However, for these market-based mechanisms to bring sustainable solutions, the states will have to increase their cooperation and revise important parts of the Law of the River. For that matter, it will be necessary to redefine the property regime of the Colorado River water as truly common property. An important step in that direction would be the creation of a legitimate permanent decision-making body composed of representatives of each basin states (Upper and Lower Basin but also Mexico). Such an organ could be responsible for the planning of Colorado River utilization with provision for the environment. This organ could take charge of monitoring ICS trading according to negotiated water plans.

Chapter 8

Murray-Darling Basin Case Study

The Murray-Darling Basin in Australia has been experiencing a terrible drought for many years. Water levels in the Murray River are at their lowest level since records started 100 years ago. Water reforms have been undertaken at the national level since 1994, but their implementation has encountered many obstacles. One of the principal objectives of the national water reform was the development of market mechanisms. And in this process, the Murray-Darling Basin was chosen as the location for testing interstate water trading. In the particularly harsh context of the drought, it is difficult to evaluate if water markets are a good direction in which to go, but the experience of the Murray-Darling Basin is interesting to contrast with the recent developments in the Colorado River Basin. The general context and the institutional design that was adopted are very different from the one of the Colorado River Basin. Australia is trying to establish an interstate trade in water entitlements among individual right holders located in different states. As we will see, the development of such a scheme requires at least the harmonization of the definition of water entitlements in the different jurisdictions. In the case of the ICUA interstate banking scheme in the Lower Colorado Basin, the participants are not individual right holders like in Australia, but Lower Basin state agencies. The second scheme, the ICS program does not concern water entitlement per se but a new product for the basin states, i.e. Intentionally

ISBN: 9781843393610. Published by IWA Publishing, London, UK.

Created Surplus that is beneficial use water conserved through extraordinary means.

We argue here that these differences can be understood as resulting from different conceptions of the property regime of the shared watercourse. In order to demonstrate that, we now present the historical context and the recent negotiations that led to the current interstate trading scheme in the Murray-Darling Basin.

GENERAL OVERVIEW OF THE MURRAY-DARLING BASIN

The Murray-Darling Basin covers over one million square kilometres, or around fourteen percent of the area of Australia, the most arid continent on earth. The Murray-Darling Basin accounts for about seventy per cent of the total area of irrigated land in Australia. The Basin has been called 'Australia's food bowl,' as it is the principal source of Australian agricultural products (improved dairy, cotton, rice, horticulture, and viticulture).

The Murray-Darling Basin is characterized by low average rainfall and very high variability in rainfall. Within the Basin, rainfall varies from 1400 mm per year in the highlands (East coast) to 300 mm in the Northwest (Murray-Darling Basin Commission).

Much of the Basin is flat with extensive plains or low undulating areas less than 200 m above sea level. In 1967 Peter Davis described the Murray system in the following terms:

> 'the river flows northwest from the snowy mountains area across the very flat riverine plains, characterized by an array of anastomosing channels, anabranches and billabongs, to the Mallee sand plains. There, near Mildura, Victoria, it is joined by two of its major tributaries: the Murrumbidgee, paralleling it from the East, and the Darling from the Northeast. It then flows West into South Australia through a 6-mile-wide valley to Overland Corner where it turns south through a limestone gorge to Lakes Alexandrina and Albert and the sea at Goolwa, about 1600 miles [2,575 kilometres] from its sources'(Davis 1967, 649).

The Darling, its major tributary, is the longest river in Australia, flowing 2,739 kilometres from Northern New South Wales to its confluence with the Murray River at Wentworth, New South Wales. The Darling tributaries rise in the range of Southern Queensland. Thus, the whole Murray-Darling river system is one of the largest river basins in the world. The Basin covers 75% of the State of New South Wales, 56% of the State of Victoria and 15% of the State of Queensland, 8% of South Australia and the entire Australian Capital Territory (MDB Commission). For some 1,880 km of its length, the River marks the border between New South Wales and Victoria (see map below).

Figure 7 Murray-Darling Basin (Credit to Murray-Darling Basin Commission).

The average run-off of the River is highly variable from year to year. It was estimated to be in average 23,850 Gigaliters per year (19 million acre feet per year) by the National Land and Water Resources Audit (National Land and Water Resources Audit 2001, 25). In a drought crisis meeting in November 2006, David Dreverman, the general manager of the Murray-Darling Basin Commission, presented new figures. From 2001 to 2005, the average inflow of the Murray-Darling was between 4,000 and 5,000 GL per year (3.2 to 4 million acre feet). The long term average inflow was estimated to be around 11,200 GL per year (9 million acre feet per year). In 2006-2007, the river levels reached the lowest figures ever recorded, with an annual inflow of 550 GL (405,000 acre feet) (Dreverman 2006).

It should be mentioned that the River Murray has been significantly modified by a series of dams, locks and weirs. Important inter-basin transfers were also constructed to bring additional water in the basin. The Snowy River is diverted to the Murray and Murrumbidgee Rivers and the Glenelg River to the Wimmera River for a total of 1,200 GL per year (MDBC 2003).

In its natural state, the River Murray was sometimes reduced to a chain of saline waterholes during severe drought. In South Australia, sea water infiltrated upstream for a considerable distance from the mouth. Since the completion of Hume Dam in 1936, a continuous flow has been maintained throughout the length of the River Murray. Without storages and regulation, the River Murray would almost certainly have ceased to run during the droughts of 1938-39, 1944-45, 1967-68, 1982-83 and 1997-98 (Murray-Darling Basin Commission MDBC).

THE 1915 RIVER MURRAY AGREEMENT AMONG SOUTH AUSTRALIA, NEW SOUTH WALES, VICTORIA AND THE COMMONWEALTH

The Murray-Darling Basin was settled following the explorations of John Oxley, Hamilton Hume and Charles Sturt between 1817 and 1830. Britain formally claimed the Western part of Australia in 1829 but separate colonies were created from parts of New South Wales: South Australia in 1836, Victoria in 1851, and Queensland in 1859. The Northern Territory was founded in 1863 as part of the Province of South Australia. The establishment of irrigation in the basin soon became a priority for the various colonial governments as a way to encourage people to settle the new colonies. In order to facilitate the development of irrigation, the governments of New South Wales and Victoria rapidly adopted water legislation. Victoria abolished the riparian rights and 'nationalized' the waters of the state in the 1886 Irrigation Act. New South Wales adopted similar legislation ten years later in 1896 (1896 Water Act). South Australia continued with the riparian rights doctrine until 1976 (Water Resources Act) (Productivity Commission 2003).

With the adoption of new legislation and government support for the development of irrigation, conflict of interests over the uses of the Murray River arose in the 1880s. New South Wales and Victoria wanted to increase diversions for the development of irrigation which they presented as essential to their development. Downstream, South Australia – referring to the riparian doctrine - explained that it had an inherent right to the flow of the river to maintain its navigability (Smith 1931, 64). Navigability was then indeed of first importance for South Australia in order to develop trade activities along the river.

The management of the River Murray was also addressed during the negotiation of the Constitution of the Confederation in the late 19th century. Thus, Smith reported that a 1897 draft constitution contained a clause vesting in the federal legislature 'the control and regulation of the navigation of the River Murray, and the use of the waters thereof from where it first forms the boundary between Victoria and New South Wales'(Smith 1931, 64). Such a clause was rejected by the states that refused to lose decision-making power over the Murray River. The states continued their negotiations to find an agreement over the Murray River after the creation in 1901 of the Commonwealth of Australia.

After years of negotiations, an agreement was reached in 1914 between the Commonwealth, New South Wales, Victoria and South Australia. Smith mentioned that the agreement was based on the results of a joint board of engineers appointed by the states in 1913. The report clearly showed that, in that particular case, the interests of irrigation were more important than those of navigation, and that the expenditure of large sums for the maintenance of navigation would not be justified (Smith 1931, 65). The main provisions of the 1914 Agreement concern the construction and cost sharing of storage facilities on the upper Murray, the construction of twenty-six weirs and locks on the Murray between Blanchetown in South Australia and Echuca in Victoria and the construction of nine weirs and locks on the lower part of either the Darling or Murrumbidgee Rivers.

The Agreement provides rules to share Murray waters. South Australia was to receive at least enough water to fill Lake Victoria at once, and to maintain a regulated supply at Lake Victoria outlet of 114,000 acre feet per month (140 million m^3 or 140 GL) during summer months - from August to April, and of 47,000 acre feet (58 million m^3 or 58 GL) per month during winter months - from May to July. The total of these allocations per year is around 1,254 000 acre feet (1400 GL). New South Wales and Victoria had to contribute to this minimum quantity allocated to South Australia in the proportion of the total mean flow of their tributaries with half the flow of the main river at Albury (Smith 1931, 66).

Above all, the Agreement provides for the creation of a river basin commission formed of four members, one from each of the interested states and one member for the Commonwealth. The commission was formed by 1917. The main task of the commission consisted in the regulation of the Murray's flow and ensures that each of the three states, especially South Australia, received their agreed share of the Murray's water. Over the years, water quantity remained its prime task, but its powers were gradually extended, both by agreement and informal practice (Murray-Darling Basin Commission).

Although far from perfect, the River Murray Agreement, with the creation of a river basin commission among quasi-sovereign states, was then 'a pioneering document ahead of its time' (Murray-Darling River Commission).

THE 1992 MURRAY-DARLING BASIN AGREEMENT

The continued expansion of irrigation, especially after the Second World War, led to increased salinity problems in the river basin. The role of the River Murray Commission was thus progressively extended, through a series of amendments to the Agreement, to water quality issues and other environmental problems. However, with the constant increase in diversions for irrigation, it soon became necessary to adopt new institutions covering the basin as a whole and attributing more power to the commission.

The Murray-Darling Basin Agreement was signed by New South Wales, Victoria, South Australia and the Commonwealth in 1992. Queensland also became a signatory in 1996 (Schedule D) and the Australian Capital Territory in 1998 (Schedule H). In clause 1 of the Agreement, the purpose of the Agreement is defined in the following terms:

> 'to promote and co-ordinate effective planning and management for the equitable efficient and sustainable use of the water, land and other environmental resources of the Murray-Darling Basin.'

It should be underlined that this Agreement represents the largest integrated catchment management program in the world, both in area and scope. The total area of the watersheds of the Murray and Darling Rivers is over one million square kilometres.

The Agreement creates three new institutions and describes their composition, functions, powers and responsibilities: the Murray-Darling Basin Ministerial Council (decision-making body), the Murray-Darling Basin Commission (executive body), and the Community Advisory Committee (consultative body).

Up to three ministers from each government may sit on the Ministerial Council. Thus, the Council is not composed of water ministers alone but also of ministers such as ministers of land, environment or agriculture. The chair of the Ministerial Council is traditionally the Commonwealth Agriculture Minister. Decisions in the Ministerial Council are taken by consensus, which may sometimes delay important decisions.

The Commission, which is the executive body of the Murray-Darling governance system until 2008, is composed of an independent president appointed by the Ministerial Council, two Commissioners from each government and a representative from the ACT government (MDB Agreement Clause 20). Clause 10 of the Murray-Darling Basin Agreement clearly illustrates the subordination of the basin Commission to the Ministerial Council:

> 'The Ministerial Council may give directions to the Commission concerning the performance of the functions of the Commission and the exercise of its powers and the Commission shall comply with those directions'.

The main functions of the Commission are to advise the Ministerial Council in relation to the planning, development and management of the water, land and other environmental resources of the Murray-Darling Basin and to implement any measures and policy authorized by the Ministerial Council (clause 17).

The Community Advisory Committee was created to ensure public participation in the governance of the Basin. It is composed of 22 members appointed by the Ministerial Council on the basis of their skills, expertise and networks. The Community Advisory Committee has two main functions: 'to advise the Ministerial Council from a community viewpoint on critical natural resources management issues including indigenous issues within the basin'; and to disseminate and clarify decisions adopted by the Ministerial Council (MDBC).

The Murray-Darling Basin governance from 1992 to 2008 is summarized below.

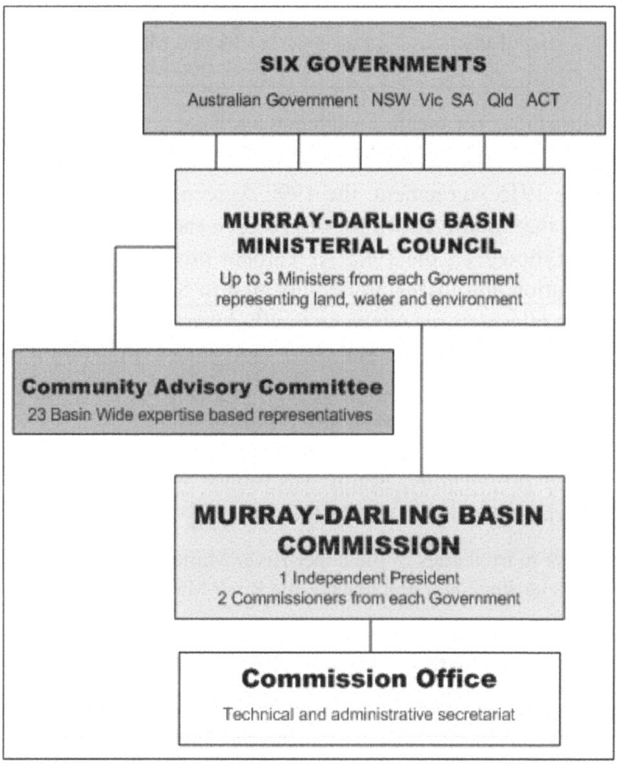

Figure 8 Murray Darling Governance 1992-2008 (Credit to Murray-Darling Basin Commission).

The 1992 Murray-Darling Basin Agreement also provides for the division of the river water among states. Thus, under Part X of the 1992 Agreement,

'South Australia is entitled to receive:

(a) the following monthly quantities of River Murray water:

Month	Quantity
July	50 500 ML
August	66 000 ML
September	77 000 ML
October	112 500 ML
November	122 000 ML
December	159 000 ML
January	159 000 ML
February	136 000 ML
March	128 000 ML
April	77 000 ML
May	35 000 ML
June	32 000 ML

(b) 58,000 megalitres per month for dilution and losses, unless the Commission determines otherwise (Part X, clause 86)'.

Thus, like the 1915 Agreement, the 1992 Agreement specifies the minimum amount of water that South Australia is to receive each month. In contrast to the 1915 Agreement, however, the 1992 Agreement provides for a special amount of water for 'dilution and losses' to be allocated to South Australia. In total, the amount of water allocated *on paper* to South Australia is higher today than in 1915. As already mentioned, in the 1915 Agreement, the minimum of water guaranteed to South Australia was around 1400 GL, this minimum is now 1850 GL if we include the amount of water provided for dilution and losses.

As in 1915, the Agreement specifies that 'New South Wales and Victoria must provide, *in equal proportions*, South Australia's apportionment'. New South Wales and Victoria are basically entitled to use:

'a) all the water in tributaries of the upper River Murray downstream of Doctors Point within its territory, before it reaches the River Murray

b) half the water in the upper River Murray upstream of Doctors Point, including any water diverted thereto by the Authority' (Part X, clause 91-93).

Contrary to the 1915 Agreement, the 1992 Agreement encompasses many other details concerning the apportionment of the waters and provides significant powers to the river basin Commission which is in charge of controlling and managing water allocations among the states (clause 95-101 'Control by Commission').

THE 1994 NATIONAL WATER REFORM FRAMEWORK

The wider context of the adoption of the 1992 Murray-Darling Basin Agreement is the discussion at the federal level of new rules for water management to improve coordination among the various states water policies. In the 1994 the Council of Australian governors agreed that the state and territory governments would adopt a strategic framework for the reform of the water industry. In 1995, the Water Reform Framework was put under the umbrella of the National Competition Policy (NCP). The general goal of the National Competition Policy is to reduce barriers to trade among Commonwealth states and territories. Under the National Competition Policy, the states and territories have agreed to review their legislation in various domains in order to identify and curtail any restriction to competition and to adopt important reforms such as in the field of electricity, gas, water and road transport. In exchange for progress against the NCP and related reform obligations, the Australian government makes NCP payments to the states and territories. For that matter, regular reports assess the achievement of the various states and territories and evaluate whether the conditions are met for the receipt of payments.

The Water Reform Framework comprises overarching principles for each state and territory to adopt. These guidelines include: the proper pricing of water, the separation of resource management and regulatory roles of government from water service provision, the implementation of comprehensive systems of water entitlements separating water property rights from land, the recognition of the environment as a legitimate user of water, the trading of water entitlements, including inter-state trading, public consultation and education, etc. (1994 Water Reform Framework).

Thus, water trading across state lines is one of the main objectives of the 1994 water reform framework. In the Murray-Darling Basin, a preliminary step to such a development was the adoption of the 'Cap' in 1995 by the MDB Ministerial Council.

CAP ON DIVERSIONS IN THE MURRAY-DARLING BASIN

In order to stop the increase in the volume of water diverted from the Basin, the Ministerial Council agreed on the principle to cap water diversions and adopted a 'Cap' in June 1995 on a temporal basis before clear rules be established. A cap is a limit on the volume of water which can be diverted from the Murray-Darling River system for consumptive uses. Within the Murray-Darling Basin Commission, a working group, composed of representatives of all the parties to the Agreement, was established to develop rules for the Cap. The parties were unable, however, to resolve a series of issues concerning the establishment of the Cap within the working group, and thus, the Ministerial Council decided to appoint an 'Independent Audit Group'. The Independent Audit Group was composed of no official states representatives. The results of the 1996 IAG's report on '*Setting the Cap*' were well received by the Ministerial Council, as out of the 49 recommendations, all except 4 were approved. Thus, by unanimous decision of the Ministerial Council, the interim Cap was transformed into a permanent cap in 1997.

Rules for the Cap were formally codified in 2000 in 'Schedule F' of the 1992 Murray-Darling Basin Agreement. The Schedule on Cap on Diversions defines the Cap as the volume of water that would have been used under 'the baseline conditions' - that is with the infrastructure supplying water, the rules for allocating water, the operating efficiency of water management systems and the entitlements that existed in June 1994. Thus, the Cap is a flexible volume and may vary from year to year in response to climatic conditions. The Cap is not the volume of water that was used in 1994. The IAG reported indeed in '*Setting the Cap*' in 1996 'because irrigation demand varies with seasonal conditions, the diversions permitted under the Cap will vary from year to year' (IAG 1996, 2). The Cap is a limit on the volume of surface water which can be diverted from the basin's rivers for consumptive uses (the Cap is calculated for each river valley in each state). The establishment of the Cap meant that water for all future demand would have to be made available either by increasing water efficiency or by trading water. The market value of irrigation water entitlements almost doubled overnight (MDB Commission).

If all states agreed on the principle to set a limit on diversions from the Murray-Darling Basin for consumptive uses, the implementation of the measure was much more complicated than thought and encountered many obstacles. The establishment of the Cap indeed corresponded to a new framework for water sharing in the basin. In *Setting the Cap*, the Independent Audit Group addressed all major equity issues and attempted to find consensual solutions. In establishing the Cap, the main difficulty was differing levels of development among riparian states. The IAG report mentioned indeed:

> 'The equity positions argued by Queensland and South Australia should be seen in the context of the history and management of water resources in the states with rivers comprising the basin. Early settlement and growth, including provision of infrastructure, saw significant increases in diversion in New South Wales and Victoria to the stage where it is now recognised that these are not sustainable. By contrast Queensland's development on its basin rivers occurred relatively late and involved limited provision of public infrastructure. Queensland has argued that there is still scope for growth in diversions although there already has been

> very rapid growth in recent years. South Australia, in recognition of environmental constraints, has historically introduced water management systems that constrained consumptive use to predominantly highly efficient crops and domestic use. It has operated under a self imposed cap since 1969 and has reduced this cap twice since 1969. It has also limited access to peak flow 'sales' or off-allocation water (Independent Audit Group 1996, 11).

In addition, the report recalled that South Australia, unlike the other states in the basin, was guaranteed a minimum allocation under the 1992 Murray-Darling Basin Agreement. Thus, South Australia had to base its water policy on this fixed minimum entitlement (1,850 GL per year) even though it could receive more (average flow received: 6,280 GL per year). Thus, contrary to New South Wales' and Victoria's irrigators, South Australian irrigators have high security water entitlements (IAG 1996, 12).

The general rule for the Cap is 'the volume of water that would have been diverted under the 1993/1994 levels of development,' but for reasons of equity, as the one mentioned above, the Cap may be adjusted and negotiated almost on a case by case basis. Under Schedule F codifying Cap rules, New South Wales and Victoria water diversions are capped to the volume of water that would have been diverted under 1993/1994 levels of development but with two small allowances for Pindari Dam (NSW) and Mokoan Storage (Victoria).

For South Australia, diversions are capped at the level that enables the development of its existing high security entitlements. This represents a small increase in diversion over 1993/1994 levels of development (MDBC 2004).

Schedule F-2 mentions that the Cap for Queensland corresponds 'to the conditions set out for each river valley in the Resource Operation Plan first adopted by the government of Queensland in that river valley'. This clause indicates that the level of the Cap was to be determined separately. Indeed, as already mentioned, the development of Queensland in the Murray-Darling Basin has been much slower than the other states in the basin because of a series of circumstances, mainly the extreme natural variability of water flow in the basin rivers. A similar procedure was undertaken for the Australian Capital Territory (the Cap was to be determined according to 'the conditions proposed by the government of that Territory and determined by the Ministerial Council on the recommendation of the Commission' schedule F-2)

In its 2006-2007 annual report on Cap Implementation[1], the Independent Audit Group mentions that

---

[1] The Independent Audit Group (IAG) is in charge of conducting an annual audit of the diversion in every designated Cap valley of the Basin comparing observed diversion against annual targets determined by valley Cap models (Independent Audit Group Report on Cap Implementation).

> 'As of October 2007, Cap arrangements have still not been finalised for some rivers in Queensland and for the border rivers and intersecting streams within New South Wales. However, progress is now advanced to the stage where we feel confident that Caps can be finalised for these systems within the next 12 months (IAG 2008).'

This report explains also that ACT has made for the first time a Cap proposal for its Territory. The IAG however considered that this proposal did not meet the principles established in its 1996 *Setting the Cap Report.* The IAG made a counter proposal in the same report released in March 2008 and finally in May 2008, an agreement was reached between the ACT and the Ministerial Council fixing the first Cap for the Territory with a special provision for population growth.

The Independent Audit Group has an important role in implementing the Cap. It is in charge of conducting an annual audit of the diversion in each state in every designated Cap valley of the Basin. The IAG determines whether the Cap has been breached in any valley and reports to the Commission. In the case of a breach of the Cap, Schedule F requires that the concerned state government reports to the Ministerial Council on the reasons why the breach occurred, the actions taken, or proposed to be taken by the state, to ensure that diversions are brought back into balance with the Cap and the period within which diversion will be brought back into balance in the Cap.

Beyond the resistance of some states to cooperate with the Cap, the implementation of the Cap is difficult for many technical reasons. As early as 1996, the IAG identified major difficulties that are still being worked out ten years later such as the harmonization and standardization of data collection methods, and the clarification of what constitute a diversion. Particularly difficult to measure are the net diversions from districts which have large return flows and also the diversions into systems such as the Yanco Creek on the Murrumbidgee and the Great Anabranch on the Darling which are a mixture of artificial diversions and natural flows.

The Murray-Darling Basin Commission concerned by the quality of data submitted by the states, commissioned Marsden Jacob Associates to undertake an audit of river diversion and intervalley water trade measurement. The *Audit of Murray-Darling Basin Cap Data Management Systems* was released in January 2005. The report found major weaknesses in the system:

> 'Across the five state and territory governments, there is little or no evidence of individual quality management systems for data on diversion levels, nor is there a common agreed approach. Each state has proceeded to date on a best efforts basis. While useful in establishing the Cap and the initial reporting regimes, a best efforts basis is no longer consistent with the spirit and obligations of Schedule F (Marsden Jacob Associates 2005, 9).'

And further:

> 'States have followed their own protocols and procedures on trade which complicates the reporting of interstate and intervalley trade. With the current low volumes of permanent interstate trade discrepancies inconsistencies are manually identified and reconciled. However manual procedures are not able to effectively deal with such discrepancies and inconsistencies where larger trade volumes are involved' (Ibid, 9).

In their 2005 audit, Marsden Jacob Associates recommend a complete revision of the requirements for reporting diversion levels and trade transactions. Among other recommendations, they suggest the creation of a 'Cap Code': a set of rules for measuring, collecting and reporting diversion and intervalley trade data. To facilitate the tracking of interstate trade, they recommend that each trade transaction should be identified by a unique number and this number should be used in all trade recording and correspondence. The findings and recommendations of this Audit Report reveal how weak has been the implementation of the Cap in the first ten years.

The implementation of the Cap in each state also required the resolution of intra-state equity issues. The implementation of the Cap within a state or territory is the responsibility of the concerned government, but the IAG established a property rights' hierarchy and rules for 'sleeper and dozer allocations' to help resolve intra-state equity issues. Indeed, a 1995 audit reported that despite the important volume of diversions, only 63% of water permitted for use was utilized. The potential activation of these sleeper or dozer rights - that is, unused water allocations - represents a challenge for the observance of the Cap. In order to comply with the Cap within each state, it was important to provide rules to arbitrate among the various water entitlements. The IAG thus recommended the following hierarchy, ranked from the highest to the lowest security level:

> '1. a statutory property right to water under existing rules which has a history of use (includes the used component of some dozer allocations);
> 2. a statutory property right to water under existing rules with no history of use (includes the unused component of some dozer allocations);
> 3. a non-statutory right[2] to use water under existing rules which has a history of use (includes the used component of some dozer allocations);
> 4. a non-statutory right to use water under existing rules with no history of use (includes the unused component of some dozer allocations);
> 5. a formal promise of a right; and
> 6. no right to water, but would have been able to get one in the past' (IAG 1996, 4).

[2] Usually issued under an Act but as a licence to use rather than a full property right.

The situation concerning dozer and sleeper rights is very different from one state to another. Quiggin notes

> 'South Australia has historically adopted a conservative allocation policy, with the result that existing statutory rights can be accommodated within the Cap. In Victoria there has been modest over-allocation, in the sense that, under existing patterns of water use, not all rights could be exercised in years of very low flow. In New South Wales, by contrast, over-allocation has been routine. As a result, even category 1 rights [high security rights] cannot be fully exercised in low-flow years' (Quiggin 2001, 87).

In its 1996 report, the Independent Audit Group clearly stated that 'The Cap should restrain diversions, not development. With the Cap in place, new developments should be allowed, provided that the water for them is obtained by improving water use efficiency or by purchasing water from existing developments' (IAG 1996, 2). Thus, as early as 1996, in alignment with the Commonwealth water reform, the clear objective of the Cap was the development of interstate water trading in the Murray-Darling Basin.

Statistics show a rapid development of intra-state temporary transfers as a consequence of the introduction of the Cap. For reasons that we will present later, the other types of transfer (permanent and interstate) remained, however, relatively small.

THE PILOT PROJECT - INTERSTATE WATER TRADING IN THE MURRAY-DARLING BASIN

In order to study and develop interstate water trading, the Murray-Darling Basin Commission launched a Pilot Project in 1997 in the Mallee Region (between Nyah - downstream of Swan Hill - and the Barrages at the mouth of the Murray). The area stretches across South Australia, Victoria and New South Wales, and was chosen for the Pilot Project because of the similarity of agricultural activities (horticulture), the homogeneity of security levels of water entitlements and the relative uniformity of water prices across the three states.

Within the Pilot Project Area, water users in South Australia, Victoria and New South Wales were allowed to buy and sell water across state boundaries. The Pilot Project was limited to the permanent transfer of high security water entitlements held by private diverters (that is private high security licences in New South Wales, water licences granted under the Water Resources Act 1997 in South Australia and private diversion licences in Victoria).

In order to implement the Pilot Project, the Murray-Darling Basin Agreement of 1992 between South Australia, Victoria and New South Wales – which allocates water among the three states - was again amended with the introduction of 'Schedule E: Interstate Transfer of Water Allocations' (MDBC). In this schedule it is mentioned that state water entitlement under Part X of the Agreement may be adjusted by the Commission in order to take into account, and give effect to, water transfers at the end of each financial year. This schedule applies to water entitlements that are permanently transferred. It does not apply to temporary transfers of water allocations. The Commission will also adjust state water Caps 'to maintain the integrity of the overall water Cap for the MDB and to reflect water allocations that have been transferred interstate under the scheme.' 'The water Cap in respect of the state of origin is to be lowered, and the water Cap in respect of the state of destination is to be correspondingly raised, to ensure that the total level of offstream diversions remains within the total of the respective water Caps of the states' (1992 Murray-Darling Basin Agreement- Schedule E-5).

In the early days of the Pilot Project, 'exchange rates' on water transactions were tested. The purpose of these exchange rates was to limit the impact of interstate water transfers on other water users and the environment. Exchange rates were applied to the amount of water transferred. Different types of exchange rates were introduced.

The first category of exchange rates - Pilot Project exchange rates - depended on the direction of the water transfers. In fact, for water transfers between Victoria and New South Wales, or downstream water transfers (from New South Wales or Victoria to South Australia), the Pilot Project exchange rate was 1. This means that in these cases, the water transfers were not affected by the exchange rate. A Pilot Project exchange rate of 0,9 or a reduction of 10% was applied to every upstream transfer (from South Australia to either New South Wales or Victoria) because of the reduced security of supply upstream of the Darling River and Lake Victoria.

A second type of exchange rate – called 'Cap exchange rate'- was to be applied to the states. Contrary to the former category of exchange rates, it was not applied to individual interstate transfers but to the total entitlements of the states involved. The Murray-Darling River Basin Commission explains:

> 'In the case of a transfer of 10 ML from South Australia to either New South Wales or Victoria, the Pilot Project exchange rate would mean a purchase of 9 ML. The 9 ML would then attract the Cap exchange rate of 0.9, reducing the quantity by a further 10 percent or 0.9 ML to 8.1ML. South Australia's Cap is then reduced by 9 ML and the upper state's Cap is increased by 8.1 ML. The overall reduction of the Basin Cap by 0.9 ML protects the security of water users and provides some compensation to the environment for the reduction in flow into South Australia. The necessary adjustments are made and registered by the MDBC' (MDBC 2004).

It was thought that the states could also apply their own exchange rates. With a potential expansion of the Pilot Project, a fourth type of exchange rate was discussed. These exchange rates were designed to facilitate interstate trade between different forms of water entitlements, and were thought of as a way to allow the conversion of one type of water entitlement into another type (e.g. entitlements with different security levels).

Evaluation of the Pilot Project

The Pilot Project was subject to a special review every two years between its opening in 1997 until its closure in 2004. The first review was delivered in December 2000 (Young, McDonald, R. Stringer and H. Bjornlund 2000). It is reported that between 1997 and 2000, 51 transactions occurred across the borders within the Pilot Project area involving 9.5 GL of water, worth, in total A$9.9 million. Three transactions were worth more than A$1 million. The average price of water transaction was between A$1,000 to A$1,150 per Megaliter. More than 90% of the water transferred was moved to South Australia. The amount of water traded represented less than half of one percent of the total amount of water used in the area (Ibid., 15-16). The report also mentioned the potential negative impacts of interstate trading on river salinity. Indeed, 'Most water is being transferred to South Australian land that has not been previously irrigated, with the consequence that river salinity can be expected to increase' (Ibid., 3).

Different reasons explained why water was moved essentially to South Australia. Some were linked to the type of water legislation in South Australia. The South Australian legislation was the only legislation in the Murray-Darling Basin to allow people to hold water allocation without owning the land that is to be irrigated. Such a disposition simplifies administrative procedures, as no environmental considerations were necessary for the transfer. Besides, the price of water was higher in South Australia than in New South Wales and Victoria.

The overall small number of permanent water transactions (less than 0,5% of the total amount of water used in the pilot area) is explained by the absence of harmonization among the various state legislations. At the time of the first review, the three states had very different environmental management and planning requirements, different fee and charge structures. In other terms, the bundle of rights associated with a one Megaliter entitlement in each state was very different. Thus, the transfer procedures were very complex and time consuming. It is reported that, on average, the administrative procedure for an interstate transfer took four to five months. Besides, the two-year review report mentions that the negotiations for new water legislation in New South Wales and other potential evolutions in the pilot zone created an atmosphere of uncertainty unsuitable for permanent transfers.

The major recommendations of the report concerned the simplification of administrative procedures for interstate transfers and the homogenization of water legislation among the three states.

In early 2004, the figures from the Murray-Darling Basin Commission show that permanent interstate trade under the Pilot Project had not evolved significantly since its inception in 1997. From 1997 to 2004, the total amount of water transferred across border was 23 Gigaliters (from 1997 to 2000: 9.5 GL of water). This amount still represents less than 1% of the water used in the Pilot Project. Some reasons for this slow development have already been presented above. Resistance to interstate trading also came from irrigation districts and local communities who feared that the sale of water would increase the cost of maintaining the water delivery infrastructure for the remaining irrigators in irrigation districts, increase taxes and also reduce employment.

Since the adoption of new legislation in 2000, New South Wales started to show some concerns with regard to the 'exchange rate' system and proposed to instead use the 'tagged' method. Discussion over the 'exchange rate' and 'tagged' trading systems created ambiguities in the minds of irrigators, who preferred temporary over permanent transfers. Finally, one should keep in mind that the Pilot Project was experienced in a context of drought, a situation that is not necessarily suitable for permanent transfers.

THE 2004 NATIONAL WATER INITIATIVE (NWI)

In order to revive the national water reform launched in 1994, all governments signed the National Water Initiative Agreement at the 25 June 2004 Council of Australian Governments meeting (with the exception of Tasmania, which signed the Agreement on 3 June 2005 and Western Australia which signed the Agreement on 6 April 2006). The National Water Initiative clearly builds on the former water reform framework of 1994. The overall objective of the National Water Initiative is again to achieve a nationally compatible water market. Section 23 of the National Water Initiative (NWI) mentions:

> 'Full implementation of this Agreement will result in a nationally-compatible, market, regulatory and planning-based system of managing surface and groundwater resources for rural and urban use that optimises economic, social and environmental outcomes'.

For that matter, the National Water Initiative Agreement establishes the National Water Commission, an independent statutory authority reporting to the

Prime Minister and, on some water reform matters, through the Prime Minister to the COAG. The Commission consists of seven commissioners: four, including the chairman, nominated by the Australian government, and three nominated jointly by the states and territories. They are appointed for their expertise in water-related fields. The National Water Commission has three main functions: to assess governments' progress in implementing the NWI, to help governments implement the NWI and to administrate two programmes under the Australian Government Water Fund (recommending projects for decision by the Australian government on financial assistance from the Water Smart Australia programme and the Raising National Water Standard programme) (Thompson 2005).

In order to facilitate interstate water trading, one of the main priorities of the NWI is the adoption by each jurisdiction of the separation between 'access' and 'use' licences (NWI, §30). This means that water access entitlements should be separated from 'regulatory approvals for water use at a particular site for a particular purpose' (not tradable). Access licences are defined as follows in the NWI:

> 'A perpetual or on-going entitlement to exclusive access to a share of water from a specified consumptive pool as defined in the relevant water plan.'

The NWI also specifies the characteristics that water access entitlements should have: be exclusive, be tradable; ability to be subdivided or amalgamated; ability to be mortgaged to access finance; and be recorded in public water registers (NWI, §31).

The NWI lays down the basis for two types of interstate water trade: trade in access entitlements (permanent) and trade in water allocation (temporary). Water allocations are defined as 'the specific volume of water allocated to water access entitlements in a given season, defined according to rules established in the relevant water plan'.

In the NWI Agreement, the governments also committed themselves to implement 'adequate measurement, monitoring and reporting systems to support public and investor confidence in the amount of water being traded, extracted for consumptive use and recovered and managed for environmental and other public benefit outcomes (NWI, §80).'

The NWI also provides for a framework to reduce uncertainty over change in water allocation. The NWI assigns risks for reduction in water allocation based on the *cause* of the reduction[3]. Thus,

---

[3] 'The Parties agree that an effective risk assignment framework occurs in the context that: the new share-based water access entitlements framework has been established; water plans have been transparently developed to determine water allocation for the entitlements; regular reporting of progress with implementing plans is occurring; and a pathway for dealing with known overallocation and/or overuse has been agreed'. (NWI,§47)

- Holders of water access entitlements bear the risk of any reduction in water allocation which arises from climatic changes or natural events (§48)
- Governments bear the risks of any reduction in water allocation which arises from changes in government policy (§50);
- Governments and water access entitlement holders share the risk of any reduction in water allocation, which arises from a change in knowledge of a water systems' capacity to sustain a particular extraction level (§49).

The principles of the water pricing reform of 1994 are recalled and further developed in the NWI as an important element for the development of interstate trade.

Finally, concerning interstate trade in the Southern Murray-Darling Basin, the governments agreed, through the adoption of the NWI, to expand the Pilot Project Area (see map below). The area over which trade can occur was extended to the entire Southern connected Murray-Darling Basin. In addition to high reliability water entitlements that could previously be traded, lower reliability products are also tradable.

Section 63 of the NWI mentions:

> 'In regard to the Southern Murray-Darling Basin, the relevant Parties (Commonwealth, New South Wales, Victoria and South Australia) that are members of the Murray-Darling Basin Ministerial Council agree to:
>
> i) take all steps necessary, including making any corresponding legislative and administrative changes, to enable *exchange rates* and/or *tagging* of water access entitlements traded from interstate sources to buyers in their jurisdictions by June 2005;
>
> ii) reduce barriers to trade in the Southern Murray-Darling Basin by taking the necessary legislative and other actions to permit open trade and ensure competitive neutrality, and to establish an interim threshold limit on the level of permanent trade out of all water irrigation areas of four per cent per annum of the total water access entitlement for the water irrigation area by June 2005'.

Once again, we observe a strong parallel between developments in the Southern Murray-Darling Basin and the wider national context. The NWI was adopted at a time when progress in the Pilot Project Area seemed to slow down, for reasons already mentioned, and more specifically because of the disagreement over exchange rates and tagging systems. It is interesting to note that this disagreement was not resolved in the NWI. The text of section 63 concerning the Southern Murray-Darling Basin leaves open these two options. The discussion over the 'tagged' and 'exchange rate' systems delayed the revision of Schedule E of the Murray-Darling Basin Agreement. After difficult negotiations, and important pressures from the Commonwealth, it was finally adopted in May 2006.

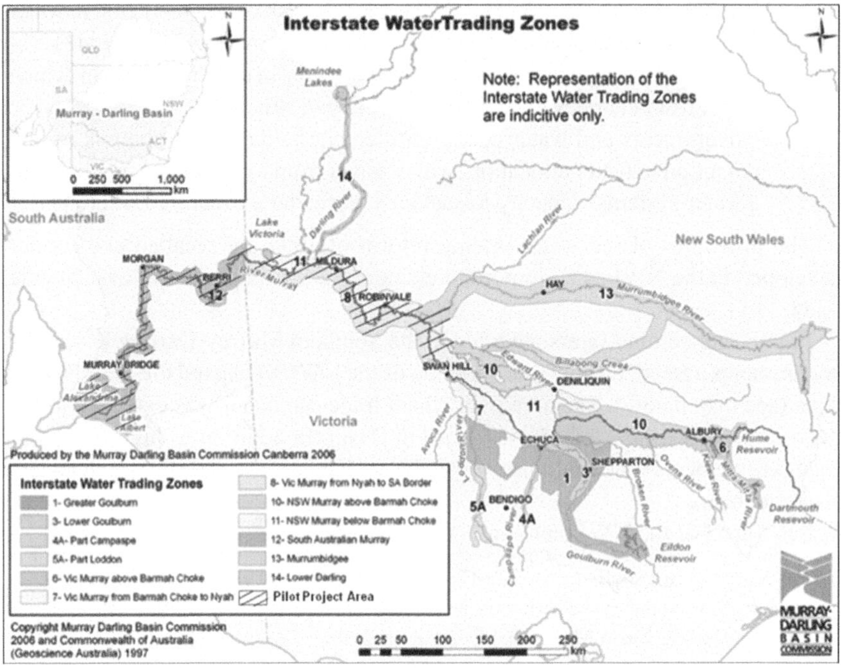

Figure 9 Interstate Trading Zones and Pilot Project Area (after MDBC).

Discussions over the exchange rates and tagging systems

In 2005, the negotiation over the trading rules for the extended trading zone remained deadlocked. Whereas the principle of exchange rates seemed to be well accepted at the inception of the Pilot Project, New South Wales, in the second half of 2005, clearly rejected the modeled exchange rate, insisting that tagging should be used for interstate trading. Victoria and South Australia preferred the exchange rate model, and even signed a bilateral agreement in early 2006 of interstate trading based on this model (February 2006).

In any case, the first assessment report of the National Water Commission[4] was released in April 2006, condemning the three Southern Murray-Darling Basin states for their failure to meet their NWI commitment by the end of 2005

[4] Previous assessments were undertaken by the National Competition Council.

(National Water Commission NWC 2006). The Commission recommended a suspended National Competition Policy payment penalty of five percent for each Southern Murray-Darling Basin state. 'The failure of Southern Murray-Darling Basin states to reach agreement on the necessary arrangements is preventing the further opening up of the interstate water trading market as required by the COAG commitments, representing a major setback to the COAG water reform process' (NWC 2006).

Presenting the progress report, Ken Matthews, Chairman and Chief Executive Officer of the National Water Commission, underlined that:

> 'Water trading is occurring in some forms, but the hard edge point I want to make today is that water trading is an absolutely essential part of the National Water Initiative, an absolutely essential part, a central element to the whole concept of the National Water Initiative, and we can't get past the objective fact that permanent trading in water isn't in place yet across the Southern Murray-Darling Basin, which is where we were to start' (Ken Matthews 2006).

Under pressure from the National Water Commission to negotiate a compromise, the three states finally agreed to implement, from July 2007, a system of tagged interstate water trading at a Murray-Darling Basin Ministerial Council meeting. Victoria and South Australia were allowed to continue interstate exchange rate entitlement transfers.

Under the tag system, water entitlements remain under the regulation of the state of origin. The entitlement remains issued by the selling state, but the water is delivered to the buying state. For example, if water is traded from the Murrumbidgee Valley in New South Wales to South Australia, then the conditions applicable to the New South Wales water would apply to its use in South Australia. Tagging means that the water entitlement carries the characteristics and reliability it had in the selling state. It is not clear at this stage how and which conditions associated with water use in each state will be applied in practice.

In the initial method, called the exchange rate method, the original entitlement was to be converted into a new entitlement issued by the buying state. The Murray-Darling Basin Commission explained how such a system could have worked:

> 'For example, purchasing a product of lower reliability such as general security water entitlements from New South Wales and applying it in South Australia will require the water product to be converted to the higher reliability South Australian water licence. To ensure that there is no overall increase in diversions, the conversion or exchange rate for this example is less than 1. To illustrate how this works - if a water user in the Goulburn Valley (Victoria) buys 10 ML of General

Security water right from a water user in the Murrumbidgee Valley (New South Wales), the Goulburn Valley water user's water right is increased by 6.349 ML' (MDBC).

As already mentioned, other types of exchange rates were discussed in the early stage of the interstate Pilot Project such as the Cap exchange rate.

In any case, these discussions over 'exchange rates' and 'tagged' systems reveal the core difficulty for the establishment of interstate trading that is the different levels of security in water entitlements among the participating states. Indeed, New South Wales, which holds the lowest security entitlements (highly variable with water availability), feared to be at a disadvantage by the exchange rates system.

Discussions over exit fees of irrigation districts

Another difficulty encountered in the development of interstate trading was the question of irrigation district financing in the case where irrigators choose to sell their water entitlements outside the district. The establishment of exit fees was proposed to limit the consequences of water transfers on originating irrigation districts. Moreover, the maximum quantity of water to be transferred from an irrigation district was limited to 4% of the total amount of water used in the irrigation district. The NWI planned to eliminate this limit in 2009. In 2010, the 4% limit was still in place in most irrigation districts.

2007 NATIONAL PLAN FOR WATER SECURITY AND THE 2007 WATER ACT: COMMONWEALTH'S INTERVENTION IN THE MURRAY-DARLING BASIN GOVERNANCE

Whereas Australia was suffering from the 'worst drought in 1,000 years'[5] in late 2006-early 2007, the Prime Minister John Howard called an emergency summit in Canberra specially focused on the situation in the Murray-Darling Basin. The Australian government once again reaffirmed its priority to accelerate the

[5] 'David Dreverman, head of the Murray-Darling River Basin Commission, said: "This is more typical of a one in a 1,000-year drought, or possibly even drier, than it is of a one in 100-year event." He added that the Murray-Darling river system, which receives 4% of Australia's water, but provides three-quarters of the water consumed nationally, was already 54% below the previous record minimum. Last month it recorded its lowest ever October flows. Inflow this year was just 5% of the average' (The Guardian, 8 November 2006).

implementation of the NWI, especially interstate water trading, which is its centrepiece. In this context, a new agreement between South Australia, New South Wales and Victoria was reached to implement interstate water trading through the tagged system from July 2007 on.

In January 2007, the Prime Minister John Howard launched the National Plan for Water Security, which eventually resulted in the adoption of the 2007 Water Act.

In his National Plan, the Prime Minister highlighted important weaknesses in Australian water management, and especially in the Murray-Darling Basin. Among other things, he criticized the decision-making procedure and the functioning of the Murray-Darling Basin Commission (MDBC). He mentioned:

> 'The decisions taken by the MDBC often reflect parochial interests and do not reflect the best interest of the Basin as a whole. Examples include:
> - Failure to align water management with the NWI in the areas of water trading, over-allocation and pricing;
> - Lack of basin-wide information has led to inefficiencies in management and decision-making. For example, a basin-wide register of water entitlements, and integrated water data systems have not been developed;
> - 12 years after introducing a Cap of water use, Queensland and the ACT ignore it and NSW is regularly in breach;
> - The MDBC has known for several years that the Cap on diversions needs to be reduced and include groundwater to be effective, but this has not been achieved;
> - Activities in one state or territory that cause problems in another can still take place without sanction, leading to redistribution of economic and environmental wealth without an overarching management framework;
> - Consensus-based decision-making of the MDBC means that difficult decisions are often avoided or delayed; and
> - Widely distributed responsibilities for the management of the basin have led to inefficiency, blame-shifting and under-resourcing by state and territory governments' (Howard 2007, 12).

Howard's plan consisted of 10 priority action points and provided that A$10 billion be spent over the next 10 years to improve water efficiency and buy back over-allocated water licences.

Prime Minister Howard conditioned that plan to the transfer of the Murray-Darling Basin governance to the federal level. From January 2007 on, the Commonwealth government tried to reach an agreement with the basin states on the transfer of powers to the federal level. The state governments, especially the Victorian government, showed important resistance to Howard's plan to create a new federal administration in charge of managing the Murray-Darling Basin.

In August 2007, the national parliament adopted the 2007 Water Act, which is a softened version of the bill because no agreement was reached with Victoria. Thus, instead of replacing the MDB Commission as provided in the bill, the 2007 Water Act creates a double management structure, leaving the Murray-Darling Basin Commission intact, but allowing the establishment of a new Commonwealth administration, the Murray-Darling Basin Authority.

Malcolm Turnbull, the then Commonwealth Minister for the Environment and Water Resources, highlighted in the water bill's second reading speech the necessity to further negotiate with the state governments even though the Commonwealth had constitutional power to legislate on the matter :

> 'The Commonwealth objective remains a comprehensive Commonwealth water law. To that end, the Commonwealth will seek to negotiate an Intergovernmental Agreement (IGA) on water with the basin states, under which they will refer their powers to underpin such comprehensive legislation within 12 months of signing'(Malcolm Turnbull 2007).

The negotiations over an Intergovernmental Agreement continued in late 2007 and 2008 after the election of the opposition's leader Kevin Rudd, who essentially followed the policy of his predecessor in this field. A key Intergovernmental Agreement on 'Murray-Darling Basin Reform' was signed by the Commonwealth, New South Wales, Victoria, South Australia, Queensland and the Australian Capital Territory on 3 July 2008. In December 2008, the Water Act 2007 was amended in order to integrate the result of the intergovernmental negotiations. The Water Amendment Act suppressed the MDBC and transferred its functions to the newly created Murray-Darling Basin Authority.

Water Act 2007

The Basin Plan and the Murray-Darling Basin Authority

The Water Act 2007 is the first water reform introduced into the Parliament of Australia in 106 years. Until 2007, the management of water resources was indeed a function of the states. But with this national law, water resources management power in the Murray-Darling Basin is transferred in large part to the federal government. The Water Act 2007, as amended, establishes a new governance system for the Murray-Darling Basin by creating a new independent expert-led institution, the Murray-Darling Basin Authority. The Authority is an Australian government agency with enforcement powers. The key role of the Authority is to prepare a Basin Plan - a legally enforceable document - for the entire basin, including Sustainable Diversion Limits (SDLs), environmental watering plan and a quality and salinity plan. This means that the Authority, after consultation with the basin states, will adopt a new limit on the amount of water than can be 'taken' from the basin water resources. Whereas the Cap was a limit fixed according to a historic level of water use (the amount of water that would have been diverted under the same conditions of development as in 1994), the new SDLs set a limit on what is 'environmentally sustainable' according to 'best available scientific knowledge' and taking into account the precautionary principle. The Sustainable Diversion Limits (SDLs) vary in terms of water volume in different years according to climatic conditions, storage levels, groundwater levels, etc. Contrary to the Cap that dealt with surface water only, the SDLs include both surface and groundwater. The SLDs are set for each water resource plan area, which are delimited according to hydrological boundaries and might therefore extend across state boundaries. This means that the Basin Plan will trigger new water sharing arrangements among states (the Murray-Darling Basin Agreement is no longer in force and is currently being revised).

The basin states will be responsible for designing water resource plans in accordance with the Basin Plan. These state-based resource plans will then need to be accredited by the Commonwealth Minister for Climate Change and Water.

While the Basin Plan, as described in the Water Act 2007, can potentially bring significant positive change in the Murray-Darling Basin, it remains to be seen how it will be put into practice. The Murray-Darling Basin Authority is in charge of preparing the Basin Plan and should release it in 2011. In this process, the basin states and the various stakeholders will be consulted but the single decision-maker on the basin plan will be the Commonwealth Water Minister (clause 3.3.1). The Ministerial Council has an advisory role in the preparation of the Plan and can refer the Plan back to the Authority for reappraisal.

The New Murray-Darling Basin Governance

In addition to the basin plan, the new Basin Authority is in charge of the functions of the former Basin Commission (executive role). The composition of the Basin Authority is, however, slightly different from the composition of the former Basin Commission. Whereas the Commission was formed by a president appointed by the Ministerial Council and two commissioners from each government, the new Basin Authority is composed of six independent experts: a chair, a chief executive (both appointed by the Commonwealth government) and four part time members (collectively appointed by the Commonwealth government and the basin states).

In its new version, the size of the Ministerial Council has been reduced. Each basin state and territory could nominate up to 3 ministers to sit at the former Council, whereas now each jurisdiction nominates one minister only. It is still chaired by the Commonwealth Minister, and the decisions are still taken by unanimity. The Ministerial Council keeps key decision-making power in matters other than the Basin Plan. 'The Ministerial Council considers and determines outcomes and objectives on *major policy issues*, that are not addressed in the Basin Plan, for the management of the basin's water and other natural resources'(clause 3.3.2, emphasis added). Further, the Agreement on Murray-Darling Basin Reform mentions that 'the parties agree that within these functions the Ministerial Council will have the *decision-making role for state water shares*, the Living Murray Initiative, and, under certain circumstances, critical human needs and natural resource management programs' (clause 3.3.3, emphasis added). In its functions, it is supported by the basin Officials Committee.

The basin Officials Committee is composed of one representative from each jurisdiction and is chaired by the Commonwealth representative. Its decisions are adopted by unanimity (clause 3.6). The basin Officials Committee acts as the executive body of the Ministerial Council. It is in charge of advising the Ministerial Council on major policy issues and of giving effect to any policy or decision of the Ministerial Council, which the Ministerial Council delegates to the basin Officials Committee (clause 3.3.13.) It is also responsible 'for resolving operational management and delivery inconsistencies that arise between the application of the Basin Plan and states' management and delivery of their water entitlements and allocations including state water shares,' but if the Committee is unsuccessful in resolving the conflict, it is then referred to the Ministerial Council for strategic direction (clause 3.3.13 and clause 3.3.2).

Finally, although the word 'Advisory' was erased from the name of the 'Basin Community Committee', it essentially remains an advisory body to both the Ministerial Council ('on those matters for which the Council seeks the committee's advice') and the Basin Authority. The size and composition of the Committee were slightly modified. Whereas it was previously composed of 22 members appointed by the Ministerial Council on the basis of their skills, expertise and networks, it is now composed of 16 members only, still appointed by the Ministerial Council, but there must be 'at least 8 individuals who are water users or representatives for one or more water users' (clause 3.7).

The figure below summarizes the new Murray-Darling Basin governance. It distinguishes between two types of relationships: those concerning the Basin Plan and those concerning other matters than the basin plan.

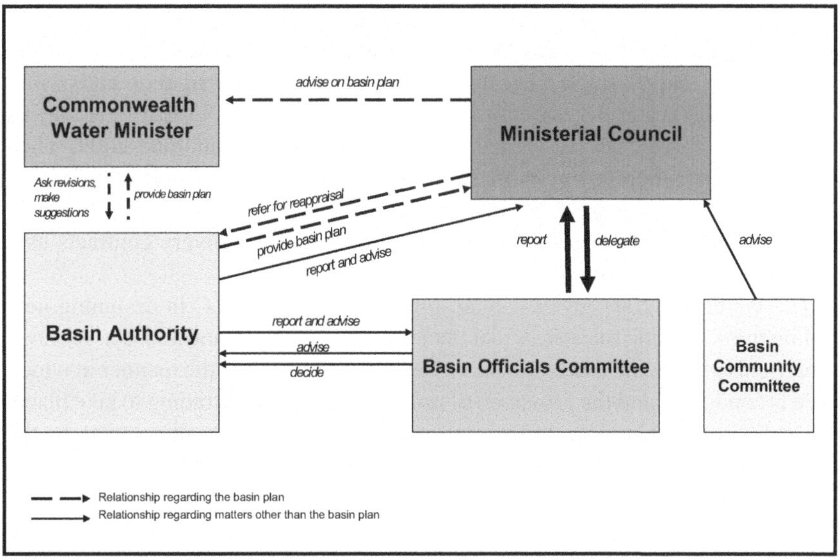

Figure 10 Murray-Darling Basin Governance 2008 (credit to Murray-Darling Basin Authority).

The new governance is clearly the result of considerable political maneuvering between the federal government and the basin states. The basin states keep important decision making powers, notably concerning 'strategic matters,' but the creation of a basin plan designed by independent experts - nominated by the Commonwealth - and adopted by the Commonwealth Water Minister is a significant step forward toward more efficient and sustainable water management in the basin.

It is too early, however, to conclude much about this new governance system. Much of the outcomes from these new institutions will depend on the personalities of the Commonwealth Water Minister, the Chair and Chief Executive of the Authority, as well as the quality and accuracy of the first Basin Plan due in 2011.

New role for the Australian Competition and Consumer Commission in preparing trading rules for the Murray-Darling Basin

According to the Water Act 2007, the new function of the Australian Competition and Consumer Commission (ACCC) is 'to ensure that water markets operate freely across state boundaries in the Murray-Darling Basin'. The Water Act 2007 provides that the ACCC is responsible for preparing 'water market and charge rules'. Water market rules relate to the transformation of irrigation rights into separately-held water access entitlements. The ACCC has already released its 'market rules,' which were signed by the Water Minister in June 2009. They require that irrigation operators must not prevent the transformation of irrigation rights into separately held, tradable water access entitlements, and that they must not demand that member irrigators terminate their delivery contracts as a condition of transformation.

The Water Act 2007 gives also an important role to ACCC in designing new trading rules. Trading rules are wider than water market rules and address 'removal of barriers to trade, the terms and processes for trading water, the manner in which trade is conducted and the provision of information to enable trading to take place' (MDBA, 2009, p13). The ACCC will provide 'advice' on trading rules to the Murray-Darling Basin Authority. These rules will be included into the Basin Plan and enforced by the ACCC. Until recently, all issues surrounding interstate water trading were dealt by the basin states. After considering submissions, including those of basin states, the ACCC released its advice to the MDBA, which was to include them into its 2010 proposed Basin Plan. The advice rules prepared by the ACCC clearly show a move toward more liberal water trading.

For the ACCC, if the water plan is well designed, and SDLs are well-established, there is no reason to have lengthy, complicated transfer approval procedures and no reason for limiting trade. Among other things, their advices include the following rules:

> -'Water trading rules should prevent specific restrictions on the ownership of water access rights based on the characteristic of a particular class of entity.'
> -'Water trading rules should specify that approval of an application to trade a water access right should not be conditional on delivery, works or use approvals, or on the holding of land where unbundling has taken place.'
> -'Water trading rules should provide that there are no restrictions on trade based on use. Environmental water holders should be treated the same as other water market participants' (ACCC 2009, p xi-xv).

The ACCC proposes keeping the 'tagged' method[6] until states' registers be inter-operational and transfer procedures be harmonized. It identifies however

[6] 'Establishing a tag is simply a method to allow water allocated to a water access entitlement to be used in a new location. The water access entitlement remains in the area of origin but its tag establishes a right to extract water at a new location (ACCC 2009, p151). Tagging provides that entitlements from each system are administered by the jurisdiction in which they were initially created, and retain all of the rights, responsibilities and risks associated with the source system.

the current administrative complexities of getting a 'tag' as a major factor in the limitation of the number of interstate trade in access entitlement (with 0 permanent trades in 2008-2009!). The ACCC recommends that the MDBA, in consultation with the basin states and other relevant stakeholders, work on improving the 'tagged' trading system and suggests alternatives such as enabling more automated water allocation trading (temporary trade) (ACCC 2009, p151).

The ACCC advises also keeping the concept of 'trading zones,' but asks that they be reassessed. The ACCC recommends: 'that the MDBA considers requiring water resource plans to:

> -Define trading zones for regulated systems, on which location-specific trading rules are referenced.
> - Where trade is restricted between two zones, the rationale behind this restriction should be explicitly stated in the water resource plan and based on physical constraints, environmental constraints, or hydrologic connections and water supply considerations' (ACCC 2009, p155).

The ACCC logically condemns the 4% limit on the permanent trade of water out of irrigation areas and recommends for their immediate removals or according to a proposed schedule (6% in 2011, 9% in 2012, 12% in 2013 and full and open trade by 2014) (ACCC 2009, p110).

Regarding the issue of trade into and out of the Murray-Darling Basin, the ACCC recommends that 'water access right trade should not be refused on the basis that the water will be used in an area outside the MDB (and the use of water inside the MDB should not be restricted solely because it was taken from a water resource outside the MDB) (ACCC 2009, p 67). The ACCC further explains 'Some stakeholders stated that water trade out of the MDB could have implications for the sustainable diversion limit. The new SDL for the MDB will reflect an environmentally sustainable level of take, whether that water will be used within the MDB or outside the MDB' (p 66). Such a rule advice will probably be subject to disagreements. Even if SDL are set up in such a way as to limit the consumptive use of water, so called 'consumptive' uses are rarely fully consumptive, and there is usually some kind of return flow. This means that from a hydrological point of view, uses in and out of the river basin contribute differently to the SDL. If such an advice rule were to be adopted, the Murray- Darling Basin Authority should fix a very strict SDL in order to carefully limit the volume of consumptive uses.

In any case, this recommendation about out of basin transfers is a good illustration of the change of level in the management of the Murray-Darling Basin. In other words, the 'public interest' becomes 'national' interest rather than regional interest.

It is important to recall that such a reform of the Murray-Darling Basin governance, and specifically the important transfer of power to the federal government, were enabled by the commitment of the federal government to spend almost 13 billion A$ over ten years to improve water management in the basin[7]. The federal government will fund various 'priority projects' such as water saving infrastructure projects. An important part of this money (3 billion A$) will be devoted to 'water buy-back' programs. '*Restoring the balance in the Murray-Darling Program*' is currently buying back water access entitlements to willing sellers across the basin. Given the way water rights have been defined in the basin, this federal water buy-back program is a necessary step in the transition toward the implementation of new sustainable diversion limits.

Conclusion

The Murray-Darling Basin has experimented with various water reforms in the last two decades. Despite these various waves of reforms, over-allocation and over-use are still a major concern in the basin. Clearly, this illustrates how difficult it is to reallocate water resources once governments have assigned too liberal water rights. The only solution today for the federal government to cope with over-allocation in the basin is to buy back water rights. Water markets are not an adequate tool for bringing back water consumption to a sustainable level. Or, more precisely, water markets should not be introduced before over-allocation and over-use are carefully addressed. In the Murray-Darling Basin, the introduction of water markets actually led to the intensification of water uses and to the reactivation of sleeper and dozer rights. The Cap on diversion established in 1996 did not significantly contribute to limit water consumption in the basin. As the Cap was designed for limiting surface water consumption only, one of its side effects was an increase in groundwater extraction. This is why the Commonwealth government is now trying to develop rules for a coordinated and integrated management of surface and groundwater resources.

Another major challenge Australia faced was to find a way to combine security in water entitlements (necessary for well-functioning water markets) with adaptability to changing climatic and hydrologic conditions. This was done in separating water access entitlements from water use licences and in defining water access entitlement as a perpetual share of a variable consumptive pool (and not a fix volume). The Water Act 2007 also provides rules to adapt to changing

[7] Specifically the Commonwealth agreed to fund 90% of the project costs of the Stage Two Food Bowl project in Victoria (up to 1 A$). This arrangement with the State of Victoria clearly helped to get the Victorian government's support for the reform.

conditions. Once the Basin Plan is adopted and new sustainable diversion limits are set, the risk of reductions in water availability will be shared among the Commonwealth and the individual rights holders according to clear principles, which can be modified according to established rules.

A core component of the Australian reform was the creation of interstate water markets in the Murray-Darling Basin. As we saw, however, the implementation of this reform has encountered many obstacles. One was the difficulty for the basin states to cooperate on the harmonization of their different water entitlements. Each jurisdiction had its own way to manage water resources and its own rules for water trading. The 'bundle of rights' linked to a water entitlement was very different from one state to another. Moreover, some states were reluctant to harmonize their water entitlements because they considered their entitlements were more secure than entitlements in other states. Indeed, because of the way the Murray-Darling was shared among different states, there is a huge difference in the security levels of water entitlements. Since South Australia was guaranteed a fixed minimum water quantity in the 1915 and 1992 Murray-Darling Basin Agreements, the water rights issued by the South Australian administration are much more secure than in the other states.

The assessment of interstate water trading activity is complicated. Interstate trading in water allocation (temporary) can be considered a success whereas trade in water access entitlement is almost inexistent.

With the federal government stepping into the management of the Murray-Darling Basin, new and uniform trading rules will be established. This will probably accelerate interstate water market activity. Interestingly, in the Australian model, the introduction of water markets is thought possible only if they are integrated into a larger river basin plan. The problem, however, is that water plans are often poorly designed because of political pressure and a lack of knowledge and understanding of water systems and their complex interconnections with the larger environment. Because the Basin Plan will be designed at the federal level by independent experts, it will probably lead to better outcomes than before. An accurate, independent and enforced water plan is a necessary condition for a water market with positive results. Could this also be possible at the international level?

In the years to come, Australia will remain an interesting laboratory to study the consequences of the introduction of interstate market mechanisms along large watercourses.

CONCLUSION

Both the Colorado River Basin and the Murray-Darling Basin have experienced a sustained drought over the last decade. Confronted with this situation, the basin states had no other choice but to acknowledge the failure of their management system and go back to the negotiation table to review their arrangements. In a general manner, these negotiations have been very difficult and lengthy. In both cases, one could measure how difficult it is to modify the definition of water rights once they have been allocated, especially if this was done 'in perpetuity' as in the US and if no provision was made for changing the rules ('rules for making rules'- Schlager and Ostrom 1992). In other words, the way the property regime of the water resources has been defined might be a major obstacle to change and to adaptive water management. In these negotiations, new tools, such as the introduction of market-based mechanisms have been discussed and tested in the two river basins studied. It should be clear by now that if we decided to focus on the Colorado River Basin and the Murray-Darling Basin it is not because they represent models for international water reform, but rather because they provide us with key elements to aid the discussion over the introduction of market-based systems along international watercourses. The comparison of these two experiences reveals very interesting differences in the institutional design and conception of a river's property regime. These two cases shed light on numerous issues including to which extent and under which conditions water markets can contribute to better water management. Major weaknesses in the design and implementation of interstate water markets have also been identified. From these analyses, interesting conclusions and policy recommendations can be drawn for international watercourse management.

The Australian model developed in the Murray-Darling Basin is in sharp contrast with the American model on the Colorado River Basin. A comparison between both basins can be drawn because they share some common physical characteristics: the areas of the Murray-Darling and the Colorado River Basins are comparable with 1,000,000 km^2 for the Murray-Darling Basin and 634,555 km^2 for the Colorado River Basin. The Murray and Colorado Rivers are respectively 2,530 and 2,352 km long. Their average runoffs are respectively 24,000 million m^3 and 18,495 million m^3, which can vary extremely from one year to another. They have both experienced extreme drought conditions in recent years, with the Colorado River's runoff being at around 5.3 m.a.f. and the Murray-Darling's runoff around 4 m.a.f. in 2004 (in 2006, it was only 405,000 a.f.). The population dependent on the Colorado River water (24 million inhabitants) is much higher than that dependent on the Murray-Darling Basin (2 million inhabitants). However, the area of irrigated land in the Murray-Darling Basin is higher than in the Colorado River Basin with respectively 3 million acres and 2 million acres. Both river basins have a dominant arid to semi-arid climate.

Both rivers flow through various states with different legal systems and water laws (4 states in the Murray-Darling Basin- plus the Australian Capital Territory;

7 states in the Colorado Basin - plus Mexico). The first agreement over the management of these shared rivers appeared in the beginning of the 20th century: in 1915 for the Murray-Darling and in 1922 for the Colorado River Basin.

At that time, however, it seems that the level of cooperation in the Murray-Darling Basin was higher than in the Colorado River Basin. The 1915 interstate agreement signed between South Australia, New South Wales, Victoria and the Council of Australian Governments provided for the establishment of a river basin commission, whereas the 1922 Colorado Compact only provided for a division of the river water into two sub-basins. No common management structure was designed in the Colorado River Basin. And still, today, no such commission for the whole Colorado Basin exists.

In a general manner, changes seem to be much more difficult to introduce in the management of the Colorado River Basin than in the Murray-Darling Basin. Indeed, the 1922 Compact allocated the whole runoff of the river in two sub-basins '*in perpetuity*' (article III). Later, the 1928 Boulder Canyon Project Act specified the allocation of each Lower Basin state and this was also done '*in perpetuity*' (section 4). The reference to 'perpetuity' clearly creates inflexibility. The Law of the River is also heavily constrained by the doctrine of prior appropriation among the different Colorado River contract holders (BCPA, section 5). According to this rule, the order of priority in case of shortage is historical. The oldest rights are the most secure entitlements, regardless of the type of uses.

In this context, there is only limited room of manoeuvre for the introduction of new management rules. Most of the basin states consider that their entitlements are their full ownership. This is why the status of apportioned but unused water remains such a conflicting issue. We saw how Arizona attempts to store an important part of its unused apportionment in underground facilities. In the Colorado Basin, many companies are investing in underground storage, which has become a major business activity.

The body of rules governing the Murray-Darling River system is very different from the one on the Colorado River. The 1915 Agreement apportioned the river water in percentages between Victoria and New South Wales. The only state that is guaranteed a fix amount of water is South Australia. This apportionment was not done in perpetuity. In 1992, the basin states managed to agree on an improved version of the 1915 Agreement. Among other things, the role and power of the Murray-Darling Basin Commission have been significantly increased. There is no prior apportionment rule in Australia. Besides, the basin states clearly mentioned in their state law that water ownership is vested in 'the Crown'. By contrast, in Western USA, the status of the Colorado water is not clearly defined.

In each river basin studied, two types of market mechanisms have been introduced. In the Colorado River basin, the first interstate banking scheme was established in 1999 (IUCA). In 2007, the basin states adopted a second banking scheme that could potentially become tradable (ICS). In the Murray-Darling Basin, from 1997 to 2007, the basin states developed an interstate trading system, but in 2007, the Commonwealth step in to design an independent Basin Plan with new trading rules.

In the Colorado River basin, the mechanisms introduced are strongly tied to the text of the Law of the River, resting on concepts such as 'unused apportionment' (ICUA) or 'surplus' (ICS). Under these circumstances, changes and evolutions remain very limited. The 1999 Federal rule on Interstate water banking allows Arizona (primarily) to store its 'intentionally created unused apportionment' in underground facilities and sell them to other lower basin states (Nevada, essentially). Arizona can also store ICUA of other Lower Basin states on their behalf in its water bank in exchange of financial compensations.

The other 'product' introduced with the adoption of the drought basin plan is Intentionally Created Surplus (ICS) that is water conserved by 'extraordinary means'. As ICS are new for all the Lower Basin states, it is easier to establish rules concerning their management. If non-Colorado contract holders such as federal agencies and NGOs are allowed to participate in the ICS program, the Law of the River will be significantly modified (although the quantity of water concerned by the ICS program is small: it is limited to 'conserved water through *extraordinary* means'). For the time being, ICS credits will only be released to the entity that created it but in the near future, it is probable that this banking scheme will involve transfers of ICS among the various users. Mexico and the US are currently negotiating over the extension of the ICS program to Mexican users and agencies. The Colorado River Basin could well soon become the first international watercourse with an international trading scheme.

In Australia, the introduction of interstate water trading was the result of the implementation of national reforms (1994 and 2004). The interstate trading scheme, regulated in schedule E of the MDB Agreement, provides that state water entitlements be modified at the end of each financial year in order to take into account permanent water transfers that occurred among individual right holders located in different states. As each state had already instituted their own water markets with their specific rules, the harmonization of the different water rights at the basin level was very difficult. In the absence of harmonization, the basin states had to agree on a method that enables entitlements originating from the different states to be traded in the same market. Interstate trading activity was very limited by the disagreement among the basin states over the 'exchange rate' system and the 'tagged' system. Denouncing the heavy and lengthy decision-making procedure of the MDB Commission, the Commonwealth government decided to take over the power of the Murray-Darling Basin in 2007 by creating a federal basin agency led by independent water experts. The new basin Authority will design a binding and enforceable Basin Plan defining new 'sustainable diversion limits'. The various entitlements originating from the different basin states will be further harmonized and unbundled. The Murray-Darling trading scheme is much more ambitious than the one in the Colorado River Basin. Indeed, the Murray-Darling Basin scheme requires important modifications in each basin state's internal laws in order to harmonize the different systems of water entitlements, pricing policies, monitoring systems and transfers reviewing procedures. It also requires significant scientific research and data collection in order to establish an accurate Basin Plan. Would such a takeover be possible in the Colorado Basin if the crisis were to worsen? At issue here is the way the property regime of the shared watercourse has been defined (at the interstate level). The definition of the shared watercourse as a common property appears more clearly in the Murray-Darling Basin context than in the Colorado River Basin context (in the texts, at least). This helps us to understand why the Murray-Darling Basin states agreed to give part of their power in terms of water management to the federal level. Given the way the Colorado River basin has been divided, it would be more surprising to observe such a transfer of power (or perhaps with significant amount of federal money).

We can summarize the different conceptions of the rivers' water property regime of the various schemes studied in the following graph[8].

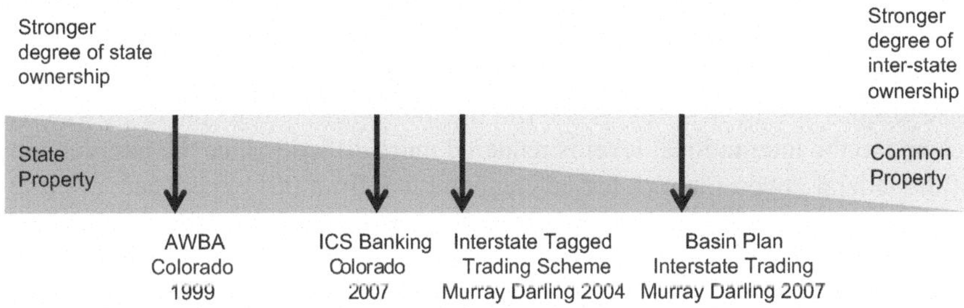

Figure 11 Classification of the Property Regimes of the Murray-Darling and Colorado Rivers' Water.

Water markets - be they local or interstate - should not be thought of as a tool to reduce water consumption. The Murray-Darling Basin states seem to have expected too much from water trading. Before introducing market mechanisms at the basin

[8] The level of analysis used to draw this comparison is the inter-state level: how the property regime of the river's water is understood by the different basin states in interstate negotiations. We are not comparing here the different water entitlements delivered to individuals within each state.

level, sustainable consumption limits should be established and enforced. We saw that this is a major challenge for both river basins. The states failed to agree and implement these limits. As a result, in 2007, the Australian Commonwealth stepped in. Until the Basin Plan is adopted, the Australian government will buy back water entitlements as a way to solve the problem of over-allocation and ease the transition toward the new SDLs. To avoid this, the basin states should adopt rules for changing the rules and assign risk in water reduction. The way the Colorado River was partitioned in 1922 in the US is probably the worst method to allocate water among riparian states because the property regime of the watercourse was not defined, no river basin commission was created, state entitlements were made in 'perpetuity', no risk in water reduction was assigned, and no rules for changing the rules were designed. Because of rapid development and climate change, the principal challenge faced by riparian states today is to design flexible allocation systems. Flexibility does not necessarily rhyme with water market. What we observed is that once water rights have been allocated and trade allowed, it is very difficult to modify these liberal water rights. Australia seems to have introduced two very interesting tools to reconcile security in water entitlements (necessary for well functioning water markets) and adaptable water management. Water entitlements have been unbundled from formal use approval and water access entitlements have been defined as a fix share of a variable consumptive pool determined by the Basin Plan. The 2007 Water Act assigns risk in water reduction between water right holders and the governments according to the cause of the reduction (drought, change in policy, evolution of scientific knowledge). The Murray-Darling Basin will certainly remain a very important case to study for the next few years. But the transposition of its interstate trading scheme to the international level is rendered more difficult since the intervention of the federal government. A lesson can be drawn from this experience: to bring sustainable outcomes, interstate water markets must be included in independent and scientific basin plan. Would that be possible at the international level? The definition of the shared watercourse as a common property- implemented by a well functioning, permanent, common decision-making body such as a river basin commission is probably a necessary but insufficient step for the design of independent and scientific basin plan.

Conclusion

Writing this book, I hope to make two contributions to the current discussions on international water governance. I tried to show that the debate over the application of international trade rules to water resources neglects a fundamental aspect of the problem: the property regime governing water resources. In this sense, the distinction between trade in water and trade in water rights is essential. I also wanted to react to some recent proposals by some economists suggesting the creation of water markets along international watercourses as solution to water conflicts. This work is a demonstration that the institutional aspects of water property regime are key elements in both discussions.

In order to address these two dimensions, I formulated the following research question: What are the consequences of the international recognition of water as an economic good on international watercourses management? This question is well founded given the fact that water is increasingly recognized as an economic good worldwide and that the concept is being implemented in various domestic reforms (such as water pricing, administrative water rights systems and water markets). We show that, to bring efficient and sustainable results, the policies aimed at the recognition of water as an economic good necessitate an integrated approach to water resources, i.e. that all water resources have to be simultaneously regulated and controlled. In this regard, underground water resources remain a key challenge for the implementation of such policies. For this reason also, the modernization of water laws and policies in many parts of the world triggers, and will increasingly trigger, the clarification of water allocations among states from

ISBN: 9781843393610. Published by IWA Publishing, London, UK.

international watercourses and aquifers. This is for example already the case in the Inkomati River Basin. However, this trend of moving away from riparianism toward more formalized systems of water rights has been accompanied at the domestic level by the incorporation of water resources into the public domain. In other words, water resources have usually been defined as common property and the success of the domestic policies recognizing water as an economic good rests on the implementation of the common property regime of these resources.

In the same manner, it is argued that the consequences of the recognition of water as an economic good on international watercourses management will depend on the type of property regime governing these particular resources. The evolution from riparianism toward water rights systems in international river basins requires the preliminary recognition of a given international watercourse as common property of all the relevant riparian states.

An important part of this work was devoted to show precisely that the property regime of international watercourses is not clearly defined and, for this reason, international trade law is increasingly interfering with the management of international watercourses. The interference of international trade law is most obvious in the emerging cases of international water transfers from international watercourses. We demonstrated that international trade law is increasingly addressed in the discussions because there is a legal and institutional vacuum.

If riparian states agreed to explicitly define their shared resource as common property and if they implemented this concept through a permanent common decision making body (such as a river basin commission), international trade law would not interfere so easily with the management of their shared watercourse. In such cases, it would be the river basin commission who would be in charge of designing rules for reviewing international water transfer proposals from the shared watercourse. If such an institutional and regulatory framework were established, trade rules would not be applicable in most cases. According to the type of property regime of the shared watercourse or aquifer, the degree of ownership of 'water rights' will vary significantly and with it the exposure to trade rules. The stronger the degree of common property of a shared watercourse, the less the delivered water rights will be exposed to trade rules. Inversely, the stronger the degree of private ownership of the water rights, the more it will be exposed to trade rules. In most cases, the transfer would not be a question of trade in water but rather a matter of water rights issuance.

The increasing reference to trade rules in water management illustrates another important reality. It is a way for non-riparian states and entities to push their interests for the resource. How and who should arbitrate the different interests in international watercourses? What is the right of non-riparian states to an international watercourse?

To answer this question, we used the concept of 'web of interests' that brings out some interesting elements (Arnold 2002). If, according to Arnold's definition, we place the object of the property at the centre of the web (that is, water resources), it follows then that the extension of the web of interests should be delimited by the natural characteristics of the resource. In other words, the 'public interest' in water resources should neither be limited by jurisdictional boundaries nor by a strict watershed limit, but by the yardstick of the protection of the resource and its ecosystem. As long as water use does not harm the resource and its ecosystem, it should be authorized, no matter how far from the point of diversion. Contrary to trade rules, the rules to evaluate transfers do not need to be even-handed for in- and out-of-basin uses. Indeed, the impact on the ecosystem will generally be different according to the type and place of the use (quantity and quality of return flow will vary). To define the property regime of water resources in this manner enables us to answer the question concerning the right of non-riparian states to international watercourses water.

It then remains to define what an adequate level of management is. Who is entitled to make the decision? We can answer this question by further developing the metaphor of the web of interests. As a web, the property regime of an international watercourse is composed of different 'layers', or different levels of decision-making corresponding to different levels of management. When the issue is the transfer from an international watercourse toward a non-riparian state, the decision making power should belong to the riparian states in common.

However, the 'international layer' of the web of interests of international watercourses is generally lacking. We further showed that, in this context, the introduction of water markets along international watercourses will not help resolve international water disputes.

To evaluate the capacity of international water markets to improve international watercourse management, it is essential to consider institutional elements of such schemes: What would be traded? Who would participate in such a market? Which rules would regulate the market? Etc. We showed that the best way to overcome the problem of the initial allocation of water rights is to limit the degree of ownership of these rights. In short, the only way to deemphasize the problem of the initial allocation of water rights is to issue water use rights, not ownership rights. This requires the recognition and implementation of a common ownership of the shared watercourse. Such recognition would also allow a more flexible water management of the water rights systems (be it tradable or not). Indeed, if the watercourse is recognized as common property, the water rights will be more easily modifiable. Contrary to some economists' arguments, flexibility in water management will not necessarily result from the introduction of water market schemes. A truly adaptive and flexible management will depend on the institutional and regulatory framework of these schemes, notably the provision of rules for changing rules and risk assignment principles in case of water reduction. The challenge for water market institutions, however, is to find an adequate balance between adaptive management, i.e. evolving rules and definition of water rights, and security in water entitlements. In this regard, Australia introduced interesting institutional tools (water access entitlements defined as a fix share of a variable pool, determined by an evolving basin plan and risk assignment according to the cause of water reduction).

In raising the controversial question of international trade in water rights as the next step in water governance, I wanted to address two aspects: whether it is a desirable evolution in water management and then, under which conditions it would be possible. We showed in the first part that water markets are a limited policy option requiring complementary regulations for administrating the complex interactions among the different water uses and between water management and other resources. Water markets can bring interesting results, but the list of requirements is quite long. The size of the market is an important element. Indeed, water markets are more efficient when the size of the market is limited, the actors are numerous and homogenous, when there is good infrastructure to carry the water (apart from means of natural conveyance), and when the technological level allows accurate data collection, monitoring, and enforcement. Finally, we analysed some interesting institutional designs for well-functioning water markets. Although not truly international, our two federal case studies help us to identify some important obstacles linked to the establishment of interstate water rights trading schemes. From these experiences we suggested some policy recommendations for the introduction of international trade in water rights and primarily stressed the importance of the recognition and implementation of the common property of the international watercourse. It should be mentioned again that international trade in water rights is more likely to develop along international watercourses than anywhere else because these watercourses are used as natural conveyance infrastructure - similar to most water market systems. In the past, international water law was influenced by federal cases and therefore might continue to follow the evolution of federal law, i.e. trade in water rights. It is difficult, however, to generalize from these two cases as they are 'extreme' cases. Indeed, the level of development and scarcity reached in these regions is particularly high in comparison to other river basins in the world. Generally speaking, tradability in water rights is a consequence of water scarcity. Therefore, before the introduction of international trade in water rights along international watercourses, the evolution most likely to occur is the progressive creation of water rights systems (not tradable water rights) in river basins under pressure. Another alternative that might concern a larger part of international river basins is the introduction of market-based instruments for the management of salinity and other water pollution problems. Further research should focus on the analysis of water pollution permits systems at the domestic and international levels.

Water governance is a fascinating interdisciplinary field of research. There is still much to learn about water systems, their interaction with different uses and the larger environment.

Successful water governance requires close collaboration among different disciplines, including natural sciences, economics, law, engineering and political science. We hope this work was a useful contribution in that direction by building bridges across some of these disciplines.

References

Aguilera Klink, Frederico, Eduardo Perez-Moriana and Juan Sanchez-Garcia. (2000). The Social Construction of Scarcity. The Case of Water in Tenerife (Canary Islands). *Ecological Economics* **34**: 23–245.

Aguilera-Klink, Frederico and M. Sanchez Padron. (2002). *Los Mercados de Agua en Tenerife*, Bilbao: Bakeaz. (Water Markets in Tenerife)

Aguilera Klink, Frederico. (2004). '¿Mercados o Administración Pública del Agua? Algunas Cuestiones Básicas'. Paper presented at the Congreso Ibérico sobre Gestión y Planificación de Aguas, April 2004, Tortosa, Spain. (Markets or Public Administration of Water? Basic Issues) http://www.congreso.us.es/ciberico/archivos_word/FedericoAguilera.doc (accessed 25 February 2010)

Al-Assar, Isabel. (2002). Future Trading of Water as a Commodity, *Newsletter 42*. http://www.waterbank.com/Newsletters/nws42.htm (accessed 25 February 2010)

Allan, J.A. (1996). Water Use and Development in Arid Regions: Environment, Economic Development and Water Resources Politics and Policy. *Water Use and Development* **5**: 107–115.

Allan, J.A. (1997). Virtual Water: a Long Term Solution for Water-Short Middle Eastern Economies? *Occasional Paper* 3. London: SOAS Water Issues Group, School of Oriental and African Studies, University of London.

Anadolu Agency News. (2000) 'Iraeli Delegation Comes to Turkey to Discuss Water Purchase'. June 2000.

Anderson, L. (2002). *The Colorado River: Utah's Perspective*, Utah Division of Water Resources, Salt Lake City, Utah. http://www.water.utah.gov/interstate/thecoloradoriverart.pdf (accessed 25 February 2010)

Arizona Water Banking Authority. (2000). *AWBA 1999 Annual Report*. AWBA, Phoenix, USA http://www.azwaterbank.gov/awba/documents/1999/report99.pdf (accessed 25 February 2010)

ISBN: 9781843393610. Published by IWA Publishing, London, UK.

Arizona Water Banking Authority. (2005). *Storage Facility Inventory Update.* AWBA, Phoenix, USA http://www.azwaterbank.gov/awba/documents/2005/Final_2005_Facility_Inventory_Update.pdf (accessed 25 February 2010)

Arizona Water Banking Authority. (2007). *Interstate Water Banking Report. Fiscal Year 2007.* AWBA, Phoenix, USA http://www.azwaterbank.gov/awba/documents/2007/Final_Interstate_Accounting_Report_FY07.pdf (accessed 25 February 2010)

Arnold, Craig Anthony. (2002). The Reconstitution of Property: Property as a Web of Interests. *Harvard Environmental Law Review* **26**(2): 281–364

Asad, M., Gabriel L., Kemper, K., Simpson, L. (1997). Management of Water Resources: Bulk Water Pricing in Brazil. *World Bank Technical Paper* 432. The World Bank, Washington, D.C.

Ausness, Richard. (1982). Water Rights Legislation in the East: A Program for Reform. *William and Mary Law Review* **24**: 547–590.

Australian Consumer and Competition Commission. (2009). *Water Trading Rules- Draft Advice.* December 2009. ACCC, Canberra, Australia.

Barlow, Maude, and Tony Clarke. (2002). *Blue Gold- The fight to Stop the Corporate Theft of the World's Water.* The New Press, New York.

Bauer Carl, (2004). *Siren Song: Chilean Water Law as a Model for International Reform*, Resources For the Future Press, Washington, D.C.

Beach, Hamner, Hewitt, Kaufman, Kurki, Oppenheimer and Wolf. (2000). *Transboundary Freshwater Dispute Resolution: Theory, Practice and Annotated References.* United Nations University Press, Tokyo and New York.

Bennett, Andrew and A. L. George. (2001). Case Studies and Process Tracing in History and Political Science: Similar Strokes for Different Foci. In: *Bridges and Boundaries: Historians, Political Scientists, and the Study of International Relations,* edited by Colin Elman and Miriam Fendius Elman, pp 137–166. MIT Press, Cambridge, MA.

Berck, Peter and Jonathan Lipow. (1994). Real and Ideal Water Rights: The Prospective for Water Rights Reform in Israel, Gaza and the West Bank. *Resources and Energy Economics* **16**: 287–301.

Boisson de Chazournes and Salman M.A. Salman. (2005). *Les Ressources en Eau et le Droit International: Water Resources and International Law.* Martinus Nijhoff Publishers, Leiden, Boston.

Boyer and Rousseau. (2006). Eau, Débattre sur le Sexe des Redevances, ou Promouvoir le Principe Pollueur-Payeur. *La Lettre Eau* n° 36, Septembre 2006, pp 3–5. France Nature Environnement. (Water, To Debate over The Sex of Charges or To Promote the Polluter Pays Principle)

Brajer, Victor, Al Church, Ronald Cummings and Phillip Farah. (1989). The Strengths and Weaknesses of Water Markets as They Affect Water Scarcity and Sovereignty Interests in the West. *Natural Resources Journal* **29**(spring): 489–509.

Briscoe, John. (1996). 'Water as an Economic Good: The Idea and What it Means in Practice'. Paper Presented by World Bank Senior Advisor at the World Congress of the International Commission on Irrigation and Drainage, September 1996, Cairo, Egypt.

Briscoe, John. (2005). *India's Water Economy: Bracing for a Turbulent Future.* Draft World Bank Report. The World Bank, Washington, D.C.

Briscoe, John, Pablo Anguita Salas and Humberto Pena. (1998). Managing Water as an Economic Resource: Reflections on the Chilean Experience. *Environment Department Paper 62.* Environmental Economic Series. The World Bank, Washington, D.C.

Bromley, Daniel. (1982). Land and Water Problems: an Institutional Perspective. *American Journal of Agricultural Economics* **64** (5): 834–844.

Bromley, Daniel. (1991). *Environment and Economy, Property Rights and Public Policy*. Basil Blackwell, Cambridge, US.

Brown Weiss, Edith. (2005). Water Transfers and International Trade Law. In: *Fresh Water and International Economic Law*, edited by Brown Weiss, Laurence Boisson de Chazournes, and Nathalie Bernasconi-Osterwalder, Oxford University Press, Oxford, UK.

Bruns, Bryan and R. Meinzen-Dick. (2000). *Negotiating Water Rights*. International Food Policy Research Institute /Vistaar, New Delhi, India.

Bruns, Bryan, Claudia Ringler, and Ruth Meinzen-Dick. (2005). *Water Rights Reform: Lessons for Institutional Design*. International Food Policy Research Institute, Washington, D.C.

Bruns, Bryan. (2006). 'Reconstituting Water Rights: Pathways for Polycentric Praxis'. Paper presented at the Conference on Survival of the Commons: Mounting Challenges and New Realities, International Association for the Study of Common Property, June 19–23, 2006, Bali, Indonesia. http://www.bryanbruns.com (accessed 25 February 2010)

Bulloch, John, Darwish, Adel. (1993). *Water Wars. Coming Conflicts in the Middle East*. Victor Gollancz, London, UK.

Buck, Susan A. (1998). *The Global Commons: An Introduction*. Island Press, Washington DC.

Business News Americas. (2005). 'Chile: President Signs New Water Code'. 13 May 2005. http://www.bnamericas.com/news/waterandwaste/President_signs_new_water_code (accessed 25 February 2010)

California Institute for Federal Policy Research. (2007). Senate Hearing Looks at Water Issues Relating to Climate Change. *California Capitol Hill Bulletin* **14** (19): 6–7 http://www.calinst.org/bul2/b1419.pdf (accessed 25 February 2010)

Carey, Janis and David Sunding. (2001). Emerging Markets in Water: A Comparative Institutional Analysis of the Central Valley and Colorado-Big Thompson Projects. *Natural Resources Journal* **41**(spring): 283–328.

Chan, Arthur. (1989). To Market or not to Market: Allocation of Interstate Waters. *Natural Resources Journal* **29**: 529–547.

Chichilnisky, Graciela. (1994). North-South Trade and the Global Environment. *The American Economic Review* **84**(4): 851–874.

Clifford, Peggy, Clay Laundry, Andrea Larsen-Hayden. (2004). *Analysis of Water Banks in the Western States*. Publication 04-11-011, Washington Department of Ecology, Olympia, Washington, USA. http://www.ecy.wa.gov/pubs/0411011.pdf (accessed 25 February 2010)

Coase, R.H. (1960). The Problem of Social Cost. *The Journal of Law and Economics* **3**: 1–44.

Colby Bonnie. (1988). Economic Impacts of Water Law: State Law and Water Market Development in the Southwest. *Natural Resources Journal* **28**(fall): 721–49.

Colby Bonnie. (1995). Regulation, Imperfect Markets, and Transaction Costs: The Elusive Quest for Efficiency in Water Allocation, in: *Handbook of Environmental Economics* ed. by Daniel Bromley, pp. 475–502. Basil Blackwell, Cambridge, USA.

Colorado New Mexico, Utah and Wyoming. (2005). *Letter to Secretary Gale Norton Re Lake Powell Water Releases*. April 18, 2005.

Commonwealth of Australia. (2004). National Water Initiative

Cooley, John K., (1984). The War over Water, *Foreign Policy*, **54**: 3–26

Corker. (1983). Sporhase v. Nebraska ex rel. Douglas: Does the Dormant Commerce Clause Really Limit the Power of State to Forbid (1) the Export of Water and (2) the Creation of a Water Right for Use in Another State? *University of Colorado Law Review* **54**: 393–445.

Cossy, Mireille. (2005). Le Statut de l'Eau en Droit International Economique. Principaux Aspects au regard des Règles de l'Organisation Mondiale du Commerce, in: *Les Ressources en Eau et le Droit International*, ed. by L. Boisson de Chazournes and Salman, pp. 169–208. Martinus Nijhoff Publishers, The Hague, The Netherlands. (The Status of Water Resources in Economic International Law)

Cummins, Tim & Associates. (2007). *An Evaluation of the Interstate Water Trade Pilot Project. Final Review for the MDBC.* Tim Cummings and Associates, Rosebank, New South Wales, Australia.

Daoudy Marwa. (2005). *Water-sharing between Syria, Turkey and Iraq: Negotiation, Security and Power Asymmetry*. CNRS Editions, Paris, France.

Daguspa and Heal. (1979). *Economic Theory and Exhaustible Resources*. Cambridge University Press, Cambridge, UK.

Davis, Peter. (1967). Australian and American Water Allocation Systems Compared. *Boston College Industrial and Commercial Law Review* **9**: 647–711

Defenders of Wildlife, Environmental Defense, National Wildlife Federation, The Nature Conservancy, Pacific Institute, Sierra Club and the Sonoran Institute. (2006). *Letter to Secretary Gale Norton Re Proposal for Colorado River Operation: 'Conservation Before Shortage II Proposal'*. July 7, 2006. http://www.usbr.gov/lc/region/programs/strategies/alternatives/CBS2.pdf (accessed 25 February 2010)

De Haan, Esther. (1997). Balancing Free Trade in Water and the Protection of Water Resources in GATT in: *The Scarcity of Water – Emerging Legal and Policy Responses* ed. by Edward Brans and al. pp. 245–259. Kluwer Law International, London, UK.

Dellapenna, Joseph. (2000). The International Joint Commission and the Export of Water from the Great Lakes : What did the Report to the governors really say? *ABA Water Resources Committee's Newsletter* **4**(1).

Dellapenna, Joseph. (2001). The Customary International Law of Transboundary Fresh Waters. *International Journal of Global Environmental Issues* **1**(3/4).

Demsetz, Harold. (1967). Toward a Theory of Property Rights, *American Economic Review* **57**(2): 347–59.

Department of Water Affairs and Forestry (DWAF). (2010). 'Free Basic Water: National View'. Department of Water Affairs and Forestry, South Africa. http://www.dwaf.gov.za/FreeBasicWater/ (accessed 25 February 2010)

Department of Water Resources of the State of California. (2005). *Colorado River Basin Climate*, Special Publication for Association of California Water Agencies and Colorado River Water Users Association Conferences, November 2005. DWR, Sacramento, Calif ornia, USA. http://www.geog.ucla.edu/downloads/634/270.pdf (accessed 25 February 2010)

Department of Water Resources of the State of California. (2008). *2009 Drought Water Bank Overview.* DWR, Sacramento, California, USA http://www.water.ca.gov/drought/docs/2009water_bank.pdf (accessed 25 February 2010)

Department of Water Resources of the State of California. (2008). *Water Transfers in 2009: Technical Guidance. Information to Parties Interested in Making Water Available to the 2009 Bank Draft and Bureau of Reclamation's Water Acquisition Program*. DWR, Sacramento, California, USA. http://www.water.ca.gov/drought/docs/TechnicalGuidancefor2009.pdf (accessed 25 February 2010)

Department of Water Resources of the State of California. (2009). Water Transfer Roundtable Discussion. Seller's Meeting. Thursday, July 2, 2009. DWR, Sacramento, California, USA. http://www.water.ca.gov/drought/docs/sellers_roundtable.pdf (accessed 25 February 2010)

De Stefano, Lucia. (2005). *Los Mercados de Aguas y la Conservacion del Medio Ambiente: Oportunidades y Retos para su Implementacion en Espana*. WWF/Adena, Madrid, Spain. (Water Markets and Environmental Conservation: Opportunities and Challenges for its Implementation in Spain)

Dendauw, I. (2000). The Great Lakes Region and Bulk Water Exports: Issues of International Trade in Water. *Water International* **25**(4): 565–571.

Dolatyar, Mostafa and Tim S. Gray. (2000). *Water Politics in the Middle East*. Saint Martin's Press, New York, USA.

Downing, Jim. (2009). 'Economic, Environmental Hurdles Block Critical California Water Transfers'. *Sacramento Bee*. May 12, 2009.

Dozier, Larry. (2005). Presentation by M. Dozier, Deputy General Manager of the Central Arizona Project at the 2005 *Natural Resources Law Centre Annual Conference on Hard Times on the Colorado River: Drought, Growth, and the Future of the Compact*. University of Colorado, Boulder, USA.

Dinar Ariel and Aaron Wolf. (1994). International Markets for Water and the Potential for Regional Cooperation: Economic and Political Perspectives in the Western Middle East. *Economic Development and Cultural Change* **43**: 4–66.

Dinar Ariel and Ashok Subramanian. (1997). Water Pricing Experiences: An International Perspective. *World Bank Technical Paper* 386. The World Bank, Washington, D.C.

Dinar, Ariel, Mark W. Rosegrant and Ruth Meinzen-Dick. (1997). Water Allocation Mechanisms: Principles and Examples. *Policy Research Working Paper* 1779. The World Bank, Washington, D.C.

Dragun, Andrew and Victor Gleeson. (1989). From Water Law to Transferability in New South Wales. *Natural Resources Journal* **29**: 645–661.

Dreverman, David. (2006). 'River Murray System: Drought Operation and Planning 2006–07 and Beyond'. Powerpoint Presentation. First Minister's Briefing November 7, 2006. http://www2.mdbc.gov.au/__data/page/54/First_Ministers_Briefing_7Nov06_MDBC.pdf (accessed 25 February 2010)

Eckstein, Gabriel. (2007). Commentary on the U.N. International Law Commission's Draft Articles on the Law of Transboundary Aquifers. *Colorado Journal of International Environmental Law and Policy* **18**: 537–610.

Elhance, Arun. (2000). *Hydropolitics in the Third World-Conflict and Cooperation in International River Basins*. United States Institute of Peace Press, Washington, D.C.

Elmusa Sharif S. (1997). *Water Conflict Economics, Politics, Law and the Palestinian-Israeli Water Resources*. Institute for Palestine Studies, Washington, D.C.

Embid, Antonio. (2002). The Evolution of Water Law and Policy in Spain. *Water Resources Development* **18**(2): 261–283.

Embid, Antonio. (2008). 'Asignación del Agua y Gestión de la Escasez en España: Los Mercados de Derechos de Aguas'. Paper presented at the Conference on Water Economics and Financing. Expo Zaragoza, Zaragoza, Spain. July 2008. (Water Allocation and Drought Management in Spain: Water Rights Water Markets)

European Union Commission. (2007). Green Paper on Market-Based Instruments for Environment and Related Policy Purposes.
Falkenmark, Malin. (1991). Environment and Development: Urgent Need for a Water Perspective. *Water International* **16**(4): 257.
Faure G.O and J.Z. Rubin. (1993). *Culture and Negotiation*. SAGE publications, London, UK.
Ferrier, Catherine. (2001). *Bottled Water: Understanding a Social Phenomenon*. World Wildlife-commissioned Study, April 2001. http://assets.panda.org/downloads/bottled_water.pdf (accessed 25 February 2010)
Fishelson, Gideon. (1994). The Water Market in Israel: An Example for increasing the Supply. *Resource and Energy Economics* **16**: 321–334.
Fisher F. (1995). The Economics of Water Dispute Resolution, Project Evaluation and Management: An Application to the Middle East. *Water Resources Development* **11**(4): 377–390.
Fisher, Franklin et al. (2005). *Liquid Assets: An Economic Approach for Water Management and Conflict Resolution in the Middle East and Beyond*. RFF Press, Washington D.C.
Gaffney, Mason. (1962). Comparison of Market Pricing and other Means of Allocating Water Resources. *Water Law and Policy in the Southeast*. Southeastern Water Law Conference, University of Georgia, November 1961, Institute of Law and Government, University of Georgia, 1962, pp. 195–229.
Gaffney, Mason. (1962). Water Law and Economic Transfers of Water: a Reply. *Journal of Financial Economics* **44**(2): 427–34.
Garduno, Hector. (2005). Lessons from Implementing Water Rights in Mexico. In: *Water Rights Reform: Lessons for Institutional Design* edited by Bruns, Bryan, Claudia Ringler, and Ruth Meinzen-Dick pp. 8–112. International Food Policy Research Institute, Washington, D.C.
Garrido, Alberto. (2003). 'Analysis of the Spanish Water Law Reform'. Paper presented at the International Working Conference on Water Rights: Institutional Options for Improving Water Allocation, February 12–15, 2003, Hanoi, Vietnam.
Gelt, Joe. (1997). Sharing Colorado River Water: History, Public Policy and the Colorado River Compact, in *Arroyo* **10**(1).
Gelt, Joe. (2004). Arizona, Nevada Are Partners in Major Water Banking Deal. *Water Resource* **13**(3).
Geneva Initiative. (2009). Annex 10- Water. Annex to the Geneva Accord. A Model Isreali-Palestinian Peace Agreement.
Girouard, Robert J. (2003). Water Export Restrictions: A Case Study of WTO Dispute Settlement Strategies and Outcomes. *Georgetown International Environmental Law Review:* **15**: 247–289.
Gleditsch, Nils Peter. (1998). Armed Conflict and the Environment: A Critique of the Literature, *Journal of Peace Research* **35**(3): 381–400.
Gleick, Peter. (1993). Water and Conflict. *International Security* **18**(1): 79–112.
Gleick, Peter (ed.). (1998). *The World's Water 1998–1999: The Biennal Report on Freshwater Resources*. Island Press, Washington, DC.
Gleick, Peter (ed.). (2000). The *World's Water 2000–2001: The Biennal Report on Freshwater Resources*. Island Press, Washington, DC.
Gleick, Peter, Garry Wolff, E. Chalecki and Rachel Reyes. (2002). Globalization and International Trade of Water in: *The World's Water 2002–2003* ed. By Peter Gleick, pp. 33–56. Island Press, Washington, DC.

Gould, George. (1989). Transfer of Water Rights. *Natural Resources Journal* **29**: 45–477.

Grant, Douglas, (1996). Commerce Clause Limits on State Regulation of Interstate Water Export, *Water Resources Update*, **105**: 10–16.

Griffin, Ronald and Fred Boadu. (1992). Water Marketing in Texas: Opportunities for Reform. *Natural Resources Journal* **32**: 265–288.

Griffin, Ronald and Gregory Characklis. (2002). Issues and Trends in Texas Water Marketing. *Water Resources Update* **121**: 29–33.

Griffin, Ronald. (2006). *Water Resource Economics: The Analysis of Scarcity, Policies, and Projects*. The MIT Press, Cambridge, USA.

Gross, Sharon. (1985). The Galloway Project and the Colorado River Compacts: Will the Compacts Bar Transbasin Water Diversions? *Natural Resources Journal* **25**: 935–960.

Güner, Serdar. (1997). The Turkish-Syrian War of Attrition: the Water Dispute. *Studies in Conflict and Terrorism* **20**: 105–116.

Hearne, R.R. and J.L. Trava. (1997). Water Markets in Mexico: Opportunities and Constraints. *Discussion Paper* 97-01. International Institute for Environment and Development, London, UK.

Haftendorn, Helga. (2000). Water and International Conflict. *Third World Quarterly* **21**(1): 51–68.

Hall Lobina and Robin de la Motte. (2004). Making Water Privatization Illegal: New Law in the Netherlands and Uruguay, *Report of the Public Services International Research Unit*, University of Greenwich, London, UK.

Hammad, Abdelazeem, (2000). 'Goal Post Changes Point to Middle East Water Conflict'. *Mideast Mirror*, June 19, 2000.

Hamner, Jesse and Aaron Wolf. (1998). Patterns in International Water Resource Treaties: The Transboundary Freshwater Dispute Database. *Colorado Journal of International Environmental Law and Policy*, 1998 Yearbook.

Hardin Garrett. (1968). The Tragedy of the Commons. *Science* **162**: 1243–1248.

Hildering Antoinette. (2005). Water as an Economic Good, in: *Les Ressources en Eau et le Droit International*, edited by L. Boisson de Chazournes and Salman, Martinus Nijhoff Publishers, The Hague, The Netherlands.

Hirschleifer, De Haven, Milliman. (1960). *Water Supply: Economics, Technology and Policy*. The University of Chicago Press, Chicago, USA.

Hoekstra, A.Y. (2003). Virtual Water Trade between Nations: A Global Mechanism Affecting Regional Water Systems. *IGBP Global Change News Letter*, **54**: 2–4.

Hodgson, Stephen. (2004). Land and Water: The Rights Interface. *FAO Legal Papers Online*. Food and Agriculture Organization of the United Nations, Roma, Italia. http://www.fao.org/legal/prs-ol/lpo36.pdf (accessed 25 February 2010)

Holden and Thobani. (1996). Tradable Water Rights: A Property Rights Approach to Resolving Water Shortages and Promoting Investment. *World Bank Policy Research Working Paper Series* 1627. The World Bank, Washington, D.C.

Homer-Dixon, Thomas. (1991). On the Threshold: Environmental Changes as Causes of Acute Conflict. *International Security* **16**(2): 76–116.

Homer-Dixon, Thomas. (1994). Environmental Scarcities and Violent Conflict: Evidence from Cases. *International Security* **19**(1): 5–40.

Howard, John. (2007). National Plan for Water Security. Prime Minister of Australia. http://pandora.nla.gov.au/pan/10052/20071016-1104/www.pm.gov.au/docs/national_plan_water_security.pdf (accessed 25 February 2010)

Howe, Charles. (2000). Protecting Public Values in a Water Market Setting: Improving Water Markets to Increase Economic Efficiency and Equity. *University of Denver Water Law Review* **3**: 357–371.

Howitt, Richard, et al. (1992). *A Retrospective on California's 1991 Emergency Drought Water Bank.* Report prepared for Department of Water Resources of the State of California.

Independent Audit Group. (1996). *Setting the Cap- Report of the Independent Audit Group.* Murray-Darling Basin Ministerial Council, Canberra, Australia http://www2.mdbc.gov.au/__data/page/86/SETTING_THE_CAP.pdf (accessed 25 February 2010)

Independent Audit Group. (2008). *Review of Cap Implementation 2006–2007- Report of the Independent Audit Group.* Murray-Darling Basin Ministerial Council, Canberra, Australia http://www2.mdbc.gov.au/__data/page/100/CAP_IAG_report_2006-07.pdf (accessed 25 February 2010)

International Centre for the Settlement of Investment Disputes. (2007). *Bayview Irrigation et al. v. United Mexican States.* Award. June 19, 2007. ICSID Case N° ARB(AF)/05/1 http://www.naftaclaims.com/Disputes/Mexico/Texas/Bayview_Jursdictional_Award_19-05-07.pdf (accessed 25 February 2010)

International Joint Commission. (1985). Great Lakes Diversions and Consumptive Uses: a Report to the Governments of the United States and Canada under the 1977 Reference. International Joint Commission, Ottawa, Ontario and Washington, D.C.

International Joint Commission. (1999). Interim Report: Protection of the Waters of the Great Lakes. International Joint Commission, Ottawa, Ontario and Washington, D.C.

International Joint Commission. (2000). Final Report: Protection of the Waters of the Great Lakes. International Joint Commission, Ottawa, Ontario and Washington, D.C. http://www.ijc.org/php/publications/html/finalreport.html

International Law Association. (2004). The Berlin Rules on Water Resources.

International Law Commission. (1994). UN Report: Commentary to Draft Articles of the UN Convention on the Law of Non-Navigational Uses of International Watercourses. http://untreaty.un.org/ilc/texts/instruments/english/commentaries/8_3_1994.pdf (accessed 25 February 2010)

Intergovernmental Panel on Climate Change. (2001). *Climate Change 2001: Impacts, Adaptation and Vulnerability.* Cambridge University Press, Cambridge, UK.

Iran Daily. (2005).'Kuwait Water accord Dead: Iran's Ministry'. 10 May 2005.

Iran Daily. (2006). 'Iran: Mineral Water Can Rival Oil in Exports'. 20 January 2006.

Iran Daily. (2007). 'Kuwait Water Accord To Be Renewed'. 26 June 2007.

Jägerskog, Anders. (2001). The Jordan River Basin: Explaining Interstate Water Cooperation through Regime Theory. *Occasional Paper* **31**: 1–7. Water Issues Study Group of the School of Oriental and African Studies, London, UK.

Jägerskog, Anders. (2002). Contributions of Regime Theory in Understanding Intersate Water Cooperation: Lessons Learned in the Jordan River Basin. In: *Hydropolitics in the Developing World: a Southern African Perspective,* edited by A.R. Turton and R. Henwood, pp. 73–80. African Water Issues Research Unit, Pretoria, South Africa.

Jerusalem Post. (2006). 'Israel and Turkey Plan Energy Pipeline'. 11 May 2006. http://new.jpost.com/Israel/Article.aspx?id=21569 (accessed 25 February 2010)

Jones, P. (2001). 'Operationalising Equitable and Reasonable Utilisation: Practice on the Columbia River'. Paper presented at the American Water Resources Association/

University of Dundee International Specialty Conference: Globalisation and Water Management: The Changing Value of Water, August 2001, Dundee, Scotland. International Water Law Research Institute, University of Dundee, UK.

Journal of Turkish Weekly. (2006). 'Turkey and Israel to Cooperate on Joint-Pipeline Project'. Friday, 12 May 2006. http://www.turkishweekly.net/news/31619/turkey-and-israel-to-cooperate-on-joint-pipeline-project.html (accessed 25 February 2010)

Kally Elisha. (1993). *Water and Peace: Water Resources and the Arab-Israeli Peace Process*. Praeger Publishers, Connecticut, USA.

Kaya, Ibrahim. (2003). *Equitable Utilization: The Law of the Non-Navigational Uses of International Watercourses*. Ashgate, Burlington, VT, USA.

Kelly, Mary. (2004). *A Powerful Thirst: Water Marketing in Texas*. Environmental Defense, Austin, Texas, USA. http://www.texaswatermatters.org/pdfs/articles/powerful_thirst.pdf (accessed 25 February 2010)

Khaleej Times. (2003). 'Iran: Deal with Kuwait on Water Pipeline'. 29 December 2003.

Klare, Michael. (2001). *Resource Wars: The New Landscape of Global Conflict*. Henry Holt and Company, NY, USA.

Kraemer, R. Andreas, Eleftheria Kampa and Eduard Interwies. (2004). The Role of Tradable Permits in Water Pollution Control, *Working Paper*, Inter-American Development Bank, Sustainable Development Department, Washington, D.C.

Landais, Emmanuelle. (2009). 'Desalination Faces 'Severe' Challenges'. *Gulf News*. November 8, 2009.

Lasky, Moses. (1929). From Prior Appropriation to Economic Distribution of Water by the State – via Irrigation Administration. *Rocky Mountain Law Review* **1**(3): 161–216.

Lasserre, Frédéric. (2005). *Transferts Massifs d'Eau: Outils de Développement ou Instruments de Pouvoir?* Presses de l'Université du Québec, Sainte Foy, Québec. (Bulk Water Transfers : Development Tools or Power Instruments?)

LeMarquand, David. (1976). Politics of International River Basin: Cooperation and Management. *Natural Resources Journal* **16**: 883–901.

Libecap, Gary. (2005). Rescuing Water Markets: Lessons from Owens Valley. *PERC Policy Series* **33**: 1–28. http://www.perc.org/pdf/ps33.pdf (accessed 25 February 2010)

Lobingier Charles. (1918). Napoleon and his Code. *Harvard Law Review* **32**: 114–134

Lochhead, James, Chad Asarch, Milos Barutciski, Patrick Monahan, Gray Taylor and Pieter Schenkkan. 1999. *Report to the Council of the Great Lakes Governors: Governing the Withdrawal of Water from the Great Lakes*.

Lochhead, James. (2000). Report to the Council of the Great Lakes: a Differing Viewpoint *ABA WaterResources Committee Newsletter* **13**(3).

Lowi Miriam. (1993). *Water and Power: The Politics of a Scarce Resource in the Jordan River Basin*. Cambridge University Press, Cambridge, UK.

Lowi Miriam and Brian R. Shaw. (2000). *Environment and Security: Discourses and Practices*. Basingstoke [etc.]: Macmillan Press.

Luterbacher Urs and Ellen Wiegandt. (2000). 'Water Control and Property Rights: An Analysis of the Middle Eastern Situation'. Paper presented at the Wengen-2000 Workshop on *International Workshop on Climate Change: Implications for the Hydrological Cycle and for Water Management*.

MacCaffrey, S. (1992). A Human Right to Water: Domestic and International Implications. *Georgetown International Environmental Law Review* **5**(1): 1–24.

MacCaffrey, S. (2001). The Contribution of the UN Convention on the Law of the Non-Navigational Uses of International Watercourses. *International Journal of Global Environmental Issues* **11**(3/4): 250–263.

MacCaffrey, S. (2001). *The Law of International Watercourses - Non-navigational Uses.* Oxford University Press, Oxford, UK.

MacDonnell L., Howe C., Miller K., Rice T., Bates S. (1994). *Using Water Banks to Promote more Flexible Water Use.* US Geological Survey Final Project. Natural Resources Law Center, Boulder, USA.

MacKeachnie, Gayle F. (1970). Toward an Economic Distribution of Water Rights. *Utah Law Review* (June) 442–460.

Maher, Tom. (2006). Email dated 26 October 2006. SNWA Resource Analyst, Las Vegas, Nevada.

Marino Manuel and Karin E. Kemper. (1999). Institutional Frameworks in Successful Water Markets : Brazil, Spain, and Colorado, USA. *World Bank Technical Paper* 427. The World Bank, Washington, D.C.

Marsden Jacob Associates. (2005). *Audit of Murray-Darling Basin Cap Data Management Systems*, prepared for the Murray-Darling Basin Commission, Camberwell, Victoria, Australia.

Marty, Frank. (2001). *Managing International Rivers-Problems, Politics and Institutions.* Peter Lang, Bern, Switzerland.

Marzulla, Nancy. (2005). Request for Arbitration 19 January 2005. http://naftaclaims.com/Disputes/Mexico/Texas/TexasClaims_NOA-19-01-05.pdf (accessed 25 February 2010)

Marzulla, Nancy. (2005). Letter to ICSID. 7 April 2005. http://naftaclaims.com/Disputes/Mexico/Texas/TexasClaims_Investors_ICSID_Letter-07-04-05.pdf (accessed 25 February 2010)

Matthews, Ken. (2006). Presentation by the Chairman and Chief Executive Officer of the National Water Commission at the 2006 Water Conference, 1-30 March 2006, Brisbane, Australia.

Matsuoka, Katsumi. (2001). 'Tradable Water in GATT/WTO Law: Need for New Legal Frameworks?' Paper presented at the American Water Resources Association/University of Dundee International Specialty Conference: Globalisation and Water Management: The Changing Value of Water, Dundee, Scotland, August 2001. International Water Law Research Institute, University of Dundee.

Mehta, Lyla. (2003). Problems of Publicness and Access Rights: Perspectives from the Water Domain. In: *Providing Global Public Goods*, edited by I. Kaul. pp. 556–576 Oxford University Press, New York, USA.

Meinzen-Dick, Ruth and Lee Ann Jackson. (1996). 'Multiple Uses, Multiple Users of Water Resources'. Paper presented at International Association for the Study of Common Property Meetings Berkeley, California, June 5–9, 1996.

Mexico. (2005). Memorial on Jurisdiction (Hugo Perezcano Díaz, Hugo). 16 April 2006. http://www.naftaclaims.com/disputes_mexico.htm (accessed 25 February 2010)

Milliman, J.W. (1959). Water Law and Private Decision-making: A Critique. *Journal of Law and Economics* **2**: 41–63.

Milner, Helen. (1991). The Assumption of Anarchy in International Relations Theory: A Critique. *Review of International Studies* **17**: 67–85.

Morris, Israel and Jay R. Lund. (1995). Recent California Water Transfers: Implications for Water Management. *Natural Resources Journal* **35**: 1–32.

Murray-Darling Basin Authority. (2009). Development of Sustainable Diversion Limits for the Murray-Darling Basin. *Issues Paper*. November 2009. MDBA, Canberra, Australia

Murray-Darling Basin Authority. (2009). The Basin Plan: A Concept Statement. June 2009. MDBA, Canberra, Australia.

Murray-Darling Basin Authority. (2009). *Murray–Darling Basin Authority Annual Report 2008–2009*. MDBA, Canberra, Australia.

Murray-Darling Basin Commission. (2003). 'Water Resources Fact Sheet'. November 2003. MDBC, Canberra, Australia. http://www2.mdbc.gov.au/__data/page/20/water_resourcesver2.pdf (accessed 25 February 2010)

Murray-Darling Basin Commission. (2004). *The Cap Brochure*. MDBC, Canberra, Australia http://www2.mdbc.gov.au/__data/page/86/cap_brochure.pdf (accessed 25 February 2010)

Murray-Darling Basin Commission. (1998). 'The Pilot Interstate Water Trading Project' MDBC, Canberra, Australia http://www2.mdbc.gov.au/nrm/water_issues/water_trade/pilot_interstate_water_trading_project (accessed 25 February 2010)

National Land and Water Resources Audit. (2001). *Australian Water Resources Assessment 2000: Surface Water and Groundwater - Availability and Quality*. http://www.anra.gov.au/topics/water/pubs/summary_reports/water_resources/pubs/water-summary.pdf (accessed 25 February 2010)

National Water Commission. (2006). *National Competition Policy Assessment of Water Reform Progress*. National Water Commission, Canberra, Australia. http://www.nwc.gov.au/www/html/511-2005-national-competition-policy-assessment-of-water-reform-progress.asp (accessed 25 February 2010)

National Water Commission. (2007). *National Water Initiative: First Biennial Assessment of Progress in Implementation*. National Water Commission, Canberra, Australia. http://www.nwc.gov.au/www/html/523-2007-biennial-assessment-of-progress-in-implementation.asp?intSiteID=1 (accessed 25 February 2010)

National Water Commission. (2009). *Australian Water Markets Report 2008–2009*. National Water Commission, Canberra, Australia.

National Water Commission. (2009). *Australian Water Reform 2009. Second Biennial Assessment of Progress in Implementation of the National Water Initiative*. National Water Commission, Canberra, Australia.

Nakayama, Mikiyasu. (2003). *International Waters in Southern Africa*. United Nations University Press, Tokyo, Japan.

Office of the Special Master. (2001). Arizona Court of Appeals Interprets A.R.S. Sections 45–173 and 45–811.01. *Arizona General Stream Adjudication Bulletin*, **2**: 11–14.

Ohlsson, Leif and Turton A.R. (1999). The Turning of a Screw: Social Resource Scarcity as a Bottle-neck in Adaptation to Water Scarcity, *occasional paper n°19*, SOAS Water Issues Group. http://www.soas.ac.uk/waterissues/papers/file38362.pdf (accessed 25 February 2010)

Ore, Diego. (2009). 'Bolivia and Chile Battle for Silala Water Rights'. Reuters. 22 September 2009. http://www.minesandcommunities.org/article.php?a=9525 (accessed 25 February 2010)

Ostrom, Elinor. (2000). Reformulating the Commons. *Swiss Political Science Review* **6** (1): 29–52.

Paquerot, Sylvie. (2002). *Le Statut des Ressources Vitales en Droit International: Essai sur le Concept de Patrimoine Commun de l'Humanité*. Bruylant, Bruxelles, Belgium.

(The Status of Vital Resources in International Law: Essay on the Concept of Common Heritage of Mankind)

Perry C.J., Michael Rock, and D. Seckler. (1997). Water as an Economic Good: A Solution, or a Problem? *Research Report 14*. International Irrigation Management Institute, Colombo, Sri Lanka.

Piper, D. (1967). *The International Law of the Great Lakes*. Duke University Press, Durham, NC, USA.

Productivity Commission. (2003). *Water Rights Arrangements in Australia and Overseas*. Commission Research Paper. Productivity Commission: Melbourne, Australia. http://www.pc.gov.au/research/commissionresearch/waterrights (accessed 25 February 2010)

PriceWaterhouseCoopers. (2006). *National Water Initiative : Water Trading Study Final Report*. Prepared for the Department of the Prime Minister and Cabinet of the Commonwealth of Australia. http://www.environment.gov.au/water/publications/action/nwi-wts-report.html (accessed 25 February 2010)

Putnam R. D. (1988). Diplomacy and Domestic Politics: the Logic of Two-Level Games. *International Organization* **42**: 427–460.

Quiggin, John. (2001). Environmental Economics and the Murray-Darling River System. *The Australian Journal of Agricultural and Resource Economics* **45**(1): 67–94.

Rake, Launce. (2004). 'Arizona Board OKs Nevada Water Deal'. *Las Vegas Sun. 14* December 2004.

Reinumägi, I. U. (1986). Diverting Water from the Great Lakes: Pulling the Plug on Canada. *Valparaiso University Law Review* **20**: 299–347.

Renshaw, E.F. (1984). Conserving Water through Pricing. *Water Resource Journal* **143**: 19–22.

Richards, Alan and Singh, Nirvikar. (2000). No Easy Exit: Property Rights, Markets, and Negotiations over Water. *Working Paper No. 463* Department of Environmental Studies and Department of Economics, University of California, Santa Cruz, USA.

Rogers, De Silva, Bhatia. (2002). Water is an Economic Good: How to use Prices to Promote Equity, Efficiency and Sustainability. *Water Policy* **4**: 1–17.

Saholiarisoa, Fanja. (2009). 'L'exportation d'eau annulée'. *L'Express de Madagascar*. 6 November 2009. (Water export cancelled)

Saliba, B.C. and D.B. Bush. (1987). *Water Markets in Theory and Practice: Market Transfers, Water Values and Public Policy*. Westview Press, Boulder, USA.

Salman, Salman M.A. and Kishor Uprety. (2002). *Conflict and Cooperation on South Asia's International Rivers: a Legal Perspective*. Kluwer Law International, The Hague, The Netherlands.

Sanchez, Roberto. (1997). Water Conflicts Between Mexico and the United States : Towards a Transboundary Regional Water Market? In: *The Scarcity of Water-Emerging Legal and Policy Responses* edited by Edward H.P. Brans and al. Kluwer Law International, The Hague, The Netherlands.

Sarwan, Suharto, Tjoek Walujo Subijanto and Charles Rodgers. (2005). Development of Water Rights in Indonesia. In: *Water Rights Reform: Lessons for Institutional Design* edited by Bruns, Bryan, Claudia Ringler, and Ruth Meinzen-Dick, 237–261. International Food Policy Research Institute, Washington, D.C.

Savarin, Charles. (2009). 'Statement on the Signing of a License in Favour of Sisserou Water Inc to Extract 3 Billion Gallons of Water per Annum for Export'. Press Statement of Savarin, Minister for Public Utilities, Energy and Ports of the Commonwealth of Dominica. 12 November 2009.

Savenije, Hubert H.G. and P. Van der Zaag. (2001). Demand Management and Water as an Economic Good – Paradigms with Pitfalls, *Value of Water Research Report Series* n°8 Institute of Hydraulic Engineering, Delft, The Netherlands.

Scanlon, Cassar, Nemes. (2003). 'Water as a Human Right?' Paper presented at the 7th International Conference on Environmental Law, June 2–5 2003, Sao Paulo, Brazil. IUCN Environmental Law Programme, Gland, Switzerland.

Schlager, Edela and E. Ostrom. (1992). Property-Rights Regimes and Natural Resources: A Conceptual Analysis. *Land Economics* **68**(2): 249–262.

Scott, Anthony and Georgina Coustalin. (1995). The Evolution of Water Rights. *Natural Resources Journal* **35**: 821–979.

Seven Basin States. (2007). *Letter to Secretary Dirk Kempthorne Re Basin States' Comments on DEIS*. 30 April 2007.

Seven Basin States. (2006). *Letter to Secretary Gale Norton Re Development of Lower Basin Sortage Guidelines: 'Basin States' proposal'*. 3 February 2006. http://www.usbr.gov/lc/region/programs/strategies/consultation/Feb06BasinStatesTransmittalLetter.pdf (accessed 25 February 2010)

Shrybman, Steven. (1999). A Legal Opinion Concerning Water Export Controls and Canadian Obligations Under NAFTA and the WTO. http://www.canadians.org/water/documents/Water_Export_Controls_Aug_99.pdf (accessed 25 February 2010)

Simpson, Larry and Klas Ringskog. (1997). Water Markets in the Americas. *Directions in Development*. The World Bank, Washington, D.C.

Smets, Henri. (2008). *De l'eau à un prix abordable-La Pratique des Etats*. Publications de l'Académie de l'Eau. http://www.academie-eau.org/IMG/pdf/eau_potable_prix_abordable.pdf (accessed 25 February 2010) (Drinking Water at an Affordable Price-State Practice)

Smith, H. A. (1931). *The Economic Uses of International Rivers*. P. S. King and sons, London, UK.

Solanes, Miguel, and Fernando Gonzalez-Villarreal.(1999). *The Dublin Principles for Water as Reflected in a Comparative Assessment of Institutional and Legal Arrangements for Integrated Water Resources Management*. Global Water Partnership, Santiago, Chile.

Southern Nevada Water Authority. (2008). *2008 Nevada Water Resource Plan*. SNWA, Las Vegas, Nevada, USA.

Spiegel, Carolin. (2005). International Water Law: The Contributions of Western United States Water Law to the United Nations Convention on the Law of the Non-Navigable Uses of International Watercourses. *Duke Journal of Comparative & International Law* **15**: 333–361.

Starr, Joyce. (1991). Water Wars. *Foreign Policy* **82:** 17–30.

Tarlock, Dan. (1985). The Law of Equitable Apportionment Revisited, Updated, and Restated. *University of Colorado Law Review* **56**: 381–411.

Tarlock, Dan. (1986). Inter and Intrastate Usage of Great Lakes Waters: A Legal Overview. *Case Western Reserve Journal of International Law* **18**: 67–108.

Tarlock Dan. (1997). Current Trends in United States Water Law and Policy: Private Property Rights, Public Interest Limitations and the Creation of Markets, In: *The Scarcity of Water: Emerging Legal and Policy Responses* edited by Brans E., De Haan E., Rinzema J., Nollkaemper A. pp. 183–195. Kluwer Law International, London, UK.

Thompson, Malcolm. (2005). 'National Water Initiative: The Economics of Water Management in Australia-An Overview'. Paper prepared by the General Manager, Water

Reform Group, National Water Commission, Australian Government for the *OECD Workshop on Agriculture and Water: Sustainability, Markets and Policies*, 14–18 November, 2005, South Australia. http://www.oecd.org/secure/docDocument/0,2827,en_21571361_34281952_35586717_1_1_1_1,00.doc (accessed 25 February 2010)

Trelease, Frank. (1954). Coordination of Riparian and Appropriative Rights to the Use of Water. *Texas Law Review* **33**: 24–69.

Trealease, Frank. (1985). State Water and State Lines: Commerce in Water Resources. *University of Colorado Law Review* **56**: 347–379.

Turnbull Malcom. (2007). 'Second Reading Speech: Water Bill'. Minister for the Environment and Water Resources, 8 August 2007.

Turkish Daily News. (2000). 'Schenker: Turkey Should be Compensated on Water'. 7 January 2000.

Turkish Daily News. (2008). 'Turkey, Israel in Joint Energy Project to Link Asia to Caspian'. 18 July 2008.

Turton, Anthony and Rolland Henwood. (2002). *Hydropolitics in the Developing World: a Southern African Perspective*. African Water Issues Research Unit, Pretoria, South Africa. http://www.internationalwaterlaw.org/articles/hydropolitics_book.pdf (accessed 25 February 2010)

Turton, Anthony. (2007). 'Links between Transboundary Groundwater and the Millennium Development Goals, Human Security, Climate Change and Environment in Africa'. Keynote lecture at the 2007 World Water Week, Stockholm, August 2007, Stockholm, Sweden.

United Nations Committee on Economic, Social and Cultural Rights. (2002). General Comments n°15 on articles 11 and 12 of the International Covenant on Economic, Social and Cultural Rights.

United Nations General Assembly. (2009). Resolution on the Law of Transboundary Aquifers. GA Res.63/124, 15 January 2009.

US Bureau of Reclamation. (2002) Storage and Interstate Release Agreement among the Secretary of the Interior; the Arizona Water Banking Authority, the Southern Nevada Water Authority and the Colorado River Commission of Nevada http://www.usbr.gov/lc/region/g4000/SIRA/finagmt.pdf (accessed 25 February 2010)

US Bureau of Reclamation. (2005). *Water 2025: Preventing Crises and Conflict in the West*- Water 2025 Status Report. Bureau of Reclamation Regional Office, Boulder City, USA http://biodiversity.ca.gov/Meetings/archive/water03/water2025.pdf (accessed 25 February 2010)

US Bureau of Reclamation. (2007). *Final Environmental Impact Statement-Colorado River Interim Guidelines for Lower Basin Shortages and Coordinated Operations for Lake Powell and Lake Mead.* Bureau of Reclamation Regional Office, Boulder City, USA http://www.usbr.gov/lc/region/programs/strategies/FEIS/index.html (accessed 25 February 2010)

US Bureau of Reclamation. (2007). *Draft Colorado River Interim Guidelines for Lower Basin Shortages and Coordinated Operations for Lake Powell and Lake Mead.* November 16, 2007. Bureau of Reclamation Regional Office, Boulder City, USA http://www.usbr.gov/lc/region/programs/strategies/UpdatedDraftGuidelines.pdf (accessed 25 February 2010)

US Bureau of Reclamation. (2007). Colorado River Interim Guidelines for Lower Basin Shortages and Coordinated Operations for Lakes Powell and Mead. Bureau of Reclamation Regional Office, Boulder City, USA http://www.usbr.gov/lc/region/programs/strategies/RecordofDecision.pdf (accessed 25 February 2010)

US Bureau of Reclamation. (2008). *Draft 2009 Annual Operating Plan for Colorado River Reservoirs*. Bureau of Reclamation Regional Office, Boulder City, USA

US Geological Survey. (2004). Climatic Fluctuations, Drought, and Flow in the Colorado River Basin. *USGS Fact Sheet* 2004–3062. http://pubs.usgs.gov/fs/2004/3062/ (accessed 25 February 2010)

Viscoli, Sergio. (1991). The Resource Conservation Group Proposal to Lease Colorado River Water. *Natural Resources Journal* **31**: 887–907.

Vorster, M.P. (1987). The Lesotho Highlands Water Projects. *South African Yearbook of International Law* **13**: 95–118.

Waterbury, John. (1997). Between Unilateralism and Comprehensive Accords: Modest Steps toward Cooperation in International River Basins. *Water Resources Development* **13**(3): 279–289.

WHO/UNICEF. (2008). *Progress on Drinking-water and Sanitation: Special Focus on Sanitation*. UNICEF, New York and WHO, Geneva.

Wiel, Samuel C. (1913). Theories of Water Law. *Harvard Law Review* **27**: 530–544.

Wiel, Samuel C. (1918). Origin and Comparative Development of the Law of Watercourses in the Common Law and in the Civil Law. *California Law Review* **6**(4): 245–267.

Wiel, Samuel C. (1919). Waters: American Law and French Authority. *Harvard Law Review* **33**(2): 133–167.

Wiel, Samuel C.(1936). Fifty Years of Water Law. *Harvard Law Review* **50**: 252–304.

Wilder, Margaret. (2005). Water, Power and Social Transformation: Neoliberal Reforms in Mexico. *Vertigo-Hors série*: L'eau en Amérique du Nord: Les Actes du Colloque Organisé par l'Institut Québécois des Hautes Etudes Internationales et de l'Observatoire de Recherches Internationales sur l'Eau. http://www. vertigo.revues.org/1925?file=1 (accessed 25 February 2010)

Williams, Stephen. (1983). Free Trade in Water Resources: Sporhase v. Nebraska ex rel Douglas. *Supreme Court Economic Review* **2**: 89–110.

Wolf, Aaron. (1999). Criteria for Equitable Allocations: The Heart of International Water Conflict. *Natural Resources Forum* **23**(1): 3–30.

Woodhouse, Phil and Julia Brown. (2004). Pioneering Redistributive Regulatory Reform. A Study of Implementation of a Catchment Management Agency for the Inkomati Water Management Area. *Working Paper 89*. Centre on Regulation and Competition, University of Manchester, UK. http://www.competition-regulation.org.uk/conferences/southafrica04/Brown%20&%20Woodhouse.pdf (accessed 25 February 2010)

World Bank.(1993). *Water Resources Management: A World Bank Policy Paper*. The World Bank, Washington D.C.

Workgroup on Water Transfer. (2002). *Water Transfer Issues in California: Final Report to the California State Water Resources Control Board*. http://www.waterrights.ca.gov/watertransfer/Final%20Report%20-%20Water%20Transfer%20Group.pdf (accessed 25 February 2010)

Wouters, Patricia. (2000). *Codification and Progressive Development of International Water Law*. Kluwer Law International, London, UK.

Wouters, Patricia. (2001). 'The Legal Response to International Water Scarcity and Water Conflicts: the UN Watercourses Convention and Beyond'. http://www.africanwater.org/pat_wouters1.htm (accessed 25 February 2010)

Yoskowitz, David W. (1999). Spot Market for Water along the Texas Rio Grande: Opportunities for Water Management. *Natural Resources Journal* **39**: 345–355.

Young, Mike, Darla Hatton MacDonald, Randy Stringer and Henning Bjornlund. (2000). *Inter-state Water Trading: A Two Year Review.* Commonwealth Scientific and Industrial Research Organisation- Land and Water, Canberra, Australia. http://www2.mdbc.gov.au/__data/page/914/watertrade_2yr.pdf (accessed 25 February 2010)

Zeitouni, Becker, and Schechter. (1994). Models of Water Market Mechanisms and an Illustrative Application to the Middle East. *Resource and Energy Economics* **16**: 303–319.

Index

ISBN: 9781843393610. Published by IWA Publishing, London, UK.

IWA Publishing's authorised EU representative for General Product Safety Regulations is Diane D'Arras, 15 rue Duret, 75116 Paris, France, e-mail: safety@iwap.co.uk.

Printed and bound by CPI Group (UK) Ltd, Croydon, CR0 4YY
12/05/2026
02108866-0002